The Mystery of the Tunguska Fireball
Revisited

Surendra Verma

This book is copyright material and must not be copied, reproduced, transferred, distributed, leased, licensed, publicly performed or used in any way except as specifically permitted in writing by the copyright holder, as allowed under the terms and conditions under which it was purchased or as strictly permitted by applicable copyright laws. Any unauthorised distribution or use of this text may be a direct infringement of the author's and publisher's rights and those responsible may be liable in law accordingly.

The first edition, *The Tunguska Fireball: Solving One of the Greatest Mysteries of the 20th Century*, was published in 2005 by Icon Books Ltd, UK, as a hardcover. The paperback edition (revised) was published in 2006 as *The Mystery of the Tunguska Fireball*.

This completely revised and updated edition (in hardcover, paperback and as an e-book) was published in 2024.

Cover photo courtesy of NASA

Copyright © 2005-2024 Surendra Verma

ISBN: 9798341373945

About the author

Surendra Verma is a journalist and author based in Melbourne, Australia, since 1970. He has published numerous popular science books in Australia, the UK and the US, and his books have been translated into 14 languages other than English.

Praise for the first edition (2005) of *The Mystery of the Tunguska Fireball*

'Part scientific detective story, part planetary science text and part physics lesson by subterfuge, Surendra Verma waltzes the reader through the process of scientific debate.'
  The Guardian, UK

'The case of the Tunguska fireball continues to intrigue all kinds of scientists, so Verma is able to use it to introduce the basics of a large number of disciplines within geology, astronomy and theoretical physics. It's somehow pleasing to realise that in this case, your guess is about as good as the experts.'
  Independent on Sunday, UK

'Readers who like scientific detective stories will find this book very appealing. The author adopts the tone of a criminal investigator.'
  Astronomy Now, UK

'Verma works through each of the possible culprits, explaining their pros and cons, and giving unusually clear descriptions of all of the scientific phenomena mentioned.'
  Cosmos, Australia

'Surendra Verma provides a guide to the competing theories, which range from crackpot to scientific.'
  The Sunday Age, Australia

'This book is an entertaining read, as well as an introductory reference for several areas of science.'
Teaching Science, Australia

'For the career conspiracy theorist, Surendra Verma's book is a must-read because of the subject matter.'
Business Times, Singapore

'In my opinion, it's the best book on Tunguska in English.'
Andrei Ol'khovatov, a renowned Russian researcher on the Tunguska event

Contents

Introduction	6
1. Fire in the Sky	8
2. The Case of a Missing Meteorite	22
3. The Tale of a Fiery Comet	53
4. Asteroids Behaving Badly	70
5. Traceless Tunguska	90
6. The Incredible Journey of a Black Hole	104
7. The Matter in Question	114
8. A Blast from Below	130
9. Opening the X-files	144
10. Tesla's Death Ray	160
11. New Fireballs in the Siberian Sky	166
12. Tunguska Revisited	170
13. Whodunit?	173
Timeline: 115 Years of an Enigma	183

Introduction

The so-called Tunguska event has been part of the folklore of science since 1927 when Leonid Kulik became the first scientist to visit the explosion site. He saw an oval plateau 70 km wide where the forest had been flattened. Trees were not uprooted: instead, they were stripped of their branches, snapped off and scattered like matchsticks pointing away from the direction of the blast.

Even after a careful search, Kulik found no crater or other evidence of impact. He searched for meteorite fragments but found nothing. As there was no impact crater and no substantial remnants, a giant meteorite could not have caused the Tunguska explosion. If it were not a meteorite, then what caused the explosion?

The search for the answer to this question has generated a Tunguska industry that has kept both scientists and charlatans busy for more than a century.

Hundreds of research articles in renowned and not-so-renowned journals prove that bright scientific minds are keen to solve the riddle. Tunguska also provides them with an opportunity to test-drive new theories: black holes, ball lightning, antimatter, mirror matter and dark matter are some of the examples.

Astronomers' recent fascination with the probability of a rogue asteroid striking our planet has also refocused scientific attention on Tunguska. In 2018, at a workshop at NASA Ames Research Center about 50 experts on impact airburst re-examined the Tunguska event and said that Tunguska was the largest cosmic impact witnessed by humans.

If they are not yet convinced, adventurous people can always make a trip to Tunguska and look for evidence of 'stones from the heavens'. For non-adventurous people, there are workshops, symposia and conferences.

Numerous books, websites, conspiracy theories, sensational TV documentaries, one episode of the popular TV series, *The X-Files* (episode

8, season 4, premiered on the Fox network on 24 November 1996), and so on prove that Tunguska adds enough exotic spices to make science palatable. For science fiction fans, Tunguska has boundless skies for spaceships to roam.

Whichever way you look, the Tunguska event is a fascinating journey in serious science, speculative science and science fiction. This book attempts to give you a glimpse of that journey.

A note on units, terms and spellings

Metric units are used in this book. Even if you are not familiar with them, this should not diminish your enjoyment of reading or your understanding of the subject matter. You may assume metres (m) to be yards, kilometres (km), miles, °C °F and tonnes tons, although the values used are accurate. The energy unit 'megaton (Mt)' is explained in its first appearance in the text.

The term 'theory' is used in a general sense and includes terms such as 'hypothesis' and 'model'.

Sorry American readers, spellings, of course, are British.

Chapter 1

A Fireball in the Sky

About 7.15 a.m., 30 June 1908. The Central Siberian Plateau near the Podkamennaya Tunguska River is a remote and empty wilderness of swamps, bogs and hilly pine and cedar forests. Not a soul in sight for scores of kilometres. The shuffle of the hoofs of reindeer grazing in the morning sun and the hum of dense swarms of mosquitoes as big as a sparrow punctuate the eerie silence. The bite of these ferocious mosquitoes is as strong as that of an alligator, earning them the name 'flying alligators'.

Suddenly a blindingly bright pillar of fire, the size of a tall office building, races across the clear blue sky. The dazzling fireball moves within a few seconds from the south-southeast to the north-northwest, leaving a thick trail of light some 800 km long. It descends slowly for a few minutes and then explodes about 8 km above the ground. The explosion lasts only a few seconds, but it is so powerful that it can be compared only with an atomic bomb—1,000 Hiroshima nuclear bombs.

The explosion flattens 2,150 square km of the mighty taiga, stripping millions of ancient trees of leaves and branches and leaving them bare like telegraph poles and scattering them like matchsticks. A dark mushroom cloud of dust rises to a height of 80 km over the area after the explosion. A black rain of debris and dirt follows. Shortly afterwards, bluish clouds of ice-coated dust grains are seen against the red sky.

When the fire came by, there was no one nearby to witness it and be amazed by it. The first to be awed by it are Mongolian herders thousands of kilometres away as the fireball races through the sky before exploding in the Siberian sky. Then the deafening roar, blinding brightness, burning heat and powerful ballistic waves of the explosion frighten Siberians in towns a long-distance away.

At Vanavara, a trading station about 70 km from the explosion site, startled citizens shield their face from the fierce hot wind. A trader, S.B. Semenov, sitting outside his house, is knocked off his chair by violent shock waves. The explosion emits so much heat that it seems to be burning his shirt.

He said later that he had only a moment to note the size of the bright blue 'tube' that covered an enormous part of the sky: 'Afterwards, it became dark, and at the same time, I felt an explosion that threw me several feet from the porch and for a moment, I lost consciousness.'

He regains consciousness to hear a tremendous sound that shakes the whole house and nearly moves it off its foundation, breaks the glass in the windows and damages his barn considerably. Earth trembles and then the sky splits apart and a hot wind, as from a cannon, blows past the houses.

Another trader, P.P. Kosolopov, who is walking outside his house, feels his ears burning. He covers them with his hands and runs into his home. Inside the house, debris starts falling from the ceiling and the door of his large stove flows out. The windowpanes break, and he hears thunder disappearing to the north. When it is quieter, he goes into the yard but sees nothing else.

Several kilometres north of Vanavara, the tents of dozens of nomads and herders, including the occupants, are blown up into the air by the shock waves that follow the explosion. They all suffer slight bruises when they fall back to the ground. An older man hits a tree and breaks his arm. Another older man dies of fright. Hundreds of reindeer belonging to four separate herds are killed as the pines and cedars around them blaze. Dense smoke envelops the forest.

Fishers repairing their rafts along the banks of the Podkamennaya Tunguska River (also known as Stony Tunguska River) are thrown into the air. Their horses stumble and fall to the ground as the shock waves after the explosion pulse through the rocks.

A farmer ploughing his hillside land about 200 km south of the explosion site hears sudden bangs, as if from gunfire. His horse falls on its knees. The fir forest around him is bent over by the wind. He seizes hold of his plough with both hands so that it is not carried away. The wind is so strong that it blows away most of the soil from the surface of the ground, and then drives a wall of water up the nearby river. A flame shoots up above the forest in the north.

About 600 km to the southwest, near Kansk, the Trans-Siberian Express, built only three years earlier, jars and shakes wildly on its tracks. Passengers are almost jolted out of their seats and frightened by the loud bursts of noise. The startled driver sees the tracks ahead rippling. He brings the train to a screeching halt. Sounds of distant thunder follow.

Villagers in Znamenskoye, 700 km from the explosion site, see bright lights in the sky. After the passage of the fireball, villagers in Achayevskoye, 1,200 km from the site, hear loud explosions, like gunfire, which continue for several minutes.

Reverberations around the globe

The explosion was registered by earthquake-measuring stations some 4,000 km away in the city of St Petersburg and 5200 km away in Jena, Germany. More distant stations around the world also record earthquake tremors. We know now that these tremors would have measured 5.0 on the Richter scale (devised in 1935) suggesting a moderate earthquake.

Disturbances in Earth's magnetic field—similar to ones produced by nuclear explosions in the atmosphere—were recorded 970 km south of the explosion site by the Irkutsk Magnetic and Meteorological Observatory. The magnetic storm lasted more than four hours. Subsequent analysis of these records showed that the epicentre of the 'earthquake' coincided with the location of the explosion (latitude 60 degrees 55 minutes north, longitude 101 degrees 57 minutes east) and confirmed the accurate time of the event (0014 GMT; 7.14 a.m. local time).

Two weeks after the explosion, the Smithsonian Astrophysical Observatory and Mount Wilson Observatory in the United States recorded a marked decrease in the air's transparency. It has been suggested that this was due to the loss of vast amounts of material from the fireball as it passed through the atmosphere. Calculations showed this loss to be several million tonnes, a hundred times more than the average yearly fall of meteorite matter on Earth.

On the evening of the explosion, bright, colourful and prolonged dusks were noticed across the Continent as far as Spain. Photographs of the sky taken on that night at the Heidelberg Astronomical Observatory were severely clouded because of the bright sky.

A Hamburg photographer who took a picture of the sky at 11 p.m. described it as 'volcanic dust,' as the memory of the 1883 Krakatoa volcano, modern history's most violent eruption, was still fresh in people's mind. After sunset in Antwerp, the northern horizon appeared to be on fire. A report in Vart Land, a Stockholm evening newspaper, described a 'strange illumination' on the night of 30 June.

The nights were also unusually bright over the British Isles. On 2 July, the London Times published a letter from Miss Katharine Stephen of Huntingdon about 'the strange light in the sky' which she and her sister had observed between midnight and 12.15 a.m. on 1 July. 'It would be interesting,' she requested, 'if anyone could explain the cause of so unusual a sight.'

The next day, a Holcombe Ingleby of Brancaster wrote about the 'curious sun effects at night', which had the appearance of a dying sunset of exquisite beauty. 'This not only lasted but actually grew both in extent and intensity till 2.30 this morning,' he continued. 'I myself was aroused from sleep at 1.15, and so strong was the light at that hour that I could read a book by it in my chamber quite comfortably.'

In the same issue, the newspaper also reported that in Dublin 'a very remarkable afterglow prolonged the daylight to such an extent that it was possible to read a newspaper in the open air'.

On 4 July, the London *Times* tried to explain 'the remarkable ruddy glows which have been seen on many nights lately'. The newspaper remarked that these glows had been seen over an area extending as far as Berlin, and pointed out that there was a considerable difference of opinion as to the nature of these glows: 'Some hold that they are auroral; their colour is quite consistent with this view ... [others hold that] the phenomenon was simply an abnormal twilight glow ... We may recall the circumstances of the wonderful glows which were seen in this country in the autumn of 1883, and which were due to the dust scattered in the upper atmosphere by the terrific outburst at Krakatoa at the end of August. Those glows had many points in common with the recent ones.'

The Times noted that 'distance is no obstacle in vast cosmical phenomena of this kind, which are absolutely world-embracing' and it was possible that the dust may have come 'from some unreported volcanic eruption in some little-known region of the world'.

On 3 July, *The New York Times* reported that 'remarkable lights were observed in the northern heavens on Tuesday and Wednesday nights, the bright diffused white and yellow illumination continuing throughout the night until it disappeared at dawn,' which the newspaper attributed to 'important changes on Sun's surface, causing electrical discharges'.

Two days later the newspaper published another report from its London correspondent. The report, 'like dawn at midnight: London sees sky blue and clouds tipped with pink at that hour', said that several nights through the week were marked by strange atmospheric effects. 'Following sunsets of exceptional beauty and twilight effects remarkable even in England,' the report continued, 'the northern sky at midnight became light blue, as if the dawn were breaking, and the clouds were touched with pink, in so marked a fashion that police headquarters was rung up by several people, who believed a big fire was raging in the north of London.'

At Edinburgh Observatory the night sky was noted as being very striking, 'practically daylight'. This practical daylight, *The Scotsman* reported, caused shadows to be cast in rooms with windows facing north. The saying 'make hay while the sun shines' took on a new meaning when farmers in the north of England worked in the fields all night getting in their hay before an impending storm broke.

The night brightness slowly diminished and disappeared after a few days. But scientists continued to speculate on the cause of these 'nocturnal glows'.

W.F. Denning, an astronomer from Bristol, wrote in the weekly journal Nature on 9 July: 'I have never seen June nights so dark, and Milky Way so gorgeously displayed in the heavens ... nor have I ever noticed the sky so bright as it appeared on the nights of June 30 and July 1. The firmament was abnormally luminous, with a very strong glow all over the north at midnight.' Bohuslav Brauner of the Bohemian University in Prague also wrote in the same issue: 'The peculiar light phenomenon at midnight on June 30 ... was also observed by me at Prague ... It is reported that magnetic disturbances were experienced on the telegraphic lines, but I saw no trace of the characteristic auroral bands or columns.'

The next week Denning wrote again, this time discounting his earlier explanation that the night glows were due to aurora borealis (displays of coloured lights in the northern skies): 'Whatever the true nature of the recent exhibition may have been, it is certain that something in the air exercised the capacity of reflection in a very high degree.'

In its August issue, the magazine of the Royal Observatory at Greenwich provided a vivid account of the night glows: 'At 9.30 p.m. at Greenwich, on June 30, the sky along the north-west and north horizon was of a brilliant red, in fact, there was what is usually termed a "brilliant sunset", the only peculiarity being that the brightness stretched more to the north than is usual, and endured, so that at one o'clock in the morning it extended well across the north of the horizon, and the northern sky above was of a brightness approaching that of the southern sky at the time of Full Moon'. The unsigned article went on to say that observations failed to give any evidence that it was an auroral display but 'the light, indeed, was sufficient to take photographs of terrestrial objects'.

An excellent photograph taken shortly before midnight with an exposure of about one minute showed the domes of the Naval College of the Royal Observatory with the College's training ship Fame in the foreground.

On the other side of the Atlantic, *Scientific American* reported on 29 August that 'sky glows,' called by some of the European astronomers, aurora displays, were now the subject of interesting discussion in astronomical circles, especially among the scientists of Europe. 'For some time, a peculiar strong orange-yellow light over the horizon, the color of which was more orange in its lower parts and more yellow in its higher parts, has been observed all over northern Europe and the United States,' the report continued. 'Clouds or spiral streams of various tints were brilliantly outlined across the sky, so luminous that few stars could be seen, and the Milky Way was hardly distinguishable.'.

The report quoted Denning and Brauner from *Nature* and added that 'both say they saw no trace of the characteristic auroral bands or columns in this phenomenon'.

The scientists at the Dublin meeting of the British Association for the Advancement of Science in September 1908 did not know that the microbarograph invented in 1903 by two prominent members, W.N. Shaw (later Sir Napier Shaw) and W.H. Dines, had recorded disturbances caused by the Tunguska explosion. The microbarograph automatically records sudden small changes in atmospheric pressure but does not show changes due to ordinary rising and falling of the barometer.

During a discussion on wave motion, as a curious example of atmospheric wave motion, Shaw presented six graphs recorded at six different locations in England on 30 June 1908 at about 0514 GMT (that is, five hours after the Tunguska explosion). Each graph showed a series of airwaves for about one hour. Each wave had four clear peaks as if there had been four disturbances in Earth's atmosphere during that period. Shaw noted that the peaks lasted for about fifteen minutes, and they were then 'violently interrupted by a sudden though slight disturbance' for a similar interval. Scientists at the meeting thought that this curious phenomenon was due to a sizeable atmospheric disturbance in some unknown part of the world.

Like the night glows, the six graphs were to remain one of the unexplained mysteries of science for two decades. No one, except some observers in Siberia, was aware that a mysterious fireball had exploded in the Siberian sky.

The British Antarctic Expedition of 1907-09, led by Sir Ernest Shackleton, was wintering at the Cape Royds Station in the Antarctic when the Tunguska fireball exploded. Did Shackleton's party observe aurora australis (coloured lights in the far southern skies) at the time of the explosion? There is no evidence of this; however, there is a record of an exceptional aurora seven hours before the explosion. Was this aurora in any way related to the fireball?

Figure 1. Six microbarographs recorded at different locations in England on 30 June 1908 at about 0514 GMT. (From F.J.W. Whipple, 'The great Siberian meteor,' in the *Quarterly Journal of the Royal Meteorological Society*, vol. 56, 1930.)

The 'tongue of flame'

The Siberian newspapers of the time were no better informed than their distant British and American counterparts, but the explosion was reported widely.

In the Irkutsk newspaper *Sibir* on 15 July, a correspondent reported that on the morning of 30 June, peasants in the village of Nizhne-Karelinskoye (about 465 km from the explosion site) saw quite high above the horizon a body shining very brightly with a bluish-white light. The body was in the form of 'a pipe,' and too bright for the naked eye. It moved vertically downwards for about ten minutes before it approached the ground and pulverised the forest. A massive cloud of black smoke was formed 'and a loud crash, not like thunder, but as if from the fall of large stones or was heard. All the buildings shook, and at the same time, a forked tongue of flame broke through the cloud'. The incident frightened the villagers, and they ran out into the street in a panic. Everyone thought that the end of the world was approaching.

At the time of the explosion, the correspondent of the *Sibir* report was in Kirensk (about 500 km away) and 'heard in the northwest what sounded like gunfire repeated at intervals at least ten times and lasting in all about fifteen minutes'. He noted that the northwest-facing windows in several houses were shattered.

The newspaper *Golos Tomska* sent a correspondent to the town of Kansk (about 635 km from the site), near which a large meteorite was rumoured to have fallen, to verify the information. On 17 July, the newspaper said that 'all the details of the fall of the meteorite should be ascribed to the overactive imagination of impressionable people'.

The newspaper, however, accepted that 'there is no doubt that a meteorite fell, probably some distance away, but its huge mass and so on are very doubtful'. On 28 July, the newspaper reported an earthquake in Kansk on the morning of 30 June. The report said that Earthquake was followed by a subterranean crash and a roar as if from distant firing. 'Doors, windows and the lamps before icons were all shaken,' the report continued. 'Five to seven minutes later a second crash followed, louder than the first, accompanied by a similar roar and followed after a brief interval by yet another crash, fainter than the first two.'

A detailed and dramatic description of the event appeared in the Krasnoyarsk newspaper *Krasnoyarets* on 26 July, from the paper's correspondent in the village of Kezhma (about 215 km away). This is the only newspaper report to come from the inhabited place nearest to the explosion site. After a detailed description of 'a subterranean shock which caused buildings to tremble,' the newspaper quoted witness reports that said that 'before the first bangs were heard a heavenly body of a fiery appearance cut across the sky from south to north ... neither its size nor shape could be made out owing to its speed and particularly its unexpectedness'.

The report went on to say that when the flying object touched the horizon, a huge flame shot up that cut the sky in two. 'The bangs were heard as soon as the "tongue of flame" disappeared. On the island near the village the horses and cows became noisy and began running wildly about ... Terrible shocks were heard coming from somewhere. They shook Earth, and their unknown source inspired a kind of superstitious terror. People were literally dumbfounded.' The report also noted that the blaze must have lasted for at least a minute because it was noticed by many peasants in the fields.

The 'tongue of flame' or the 'tongue of fire' seems a common eyewitness description of the Tunguska fireball and it appears in many Siberian newspapers of the time. A report from the village of Nizhne-Ilimskoye (about 420 km away) said that the population there and in the surrounding villages saw 'a fiery body like a beam' shoot from south to northwest before they heard the thunder. The fiery body disappeared immediately after the bang and a 'tongue of fire' appeared in its place, which was followed by smoke.

Many Russian newspapers outside Siberia also published news of the bright nights and shiny clouds seen at the time. Still, there was hardly any report on the fireball (there was no *Pravda* yet: it was started by Leon Trotsky in October 1908 in Vienna, moving to St Petersburg in 1912 and then to Moscow in 1918). However, on 4 July 1908, *The Trading Industrial Gazette* (St Petersburg) ran a short piece headlined: 'more about the fall of the meteorite: yesterday we received a telegram saying only—the noise was considerable, but no stone fell.'

Figure 2. A map of Tunguska and the surrounding areas showing the extent of various eyewitness observations. Key: 1—impact site; 2—flight path of the fireball; 3—extent of fireball visibility; 4—extent of explosion sounds; 5—track of the first Tunguska expedition. (Courtesy Marek Zbik.)

The year 1908

In 1908 Russia, ruled by Tsar Nicholas II, was a country caught in political unrest and social upheaval. The investigation of an earthquake or a meteorite in remote Siberia was the last priority of the Russian authorities in the then capital St Petersburg. The First World War and the Bolshevik Revolution were to occupy the country's attention for many years. The impact site was in Russia's most remote and rugged region and no effort was made to launch a scientific expedition there. And then the 'tongue of flame' that engulfed the beautiful Siberian taiga was forgotten.

As the bright night skies disappeared from Europe, scientists' and newspapers' attention also moved from this strange phenomenon to other scientific and technological marvels of the year. The year's rich harvest includes:

- Hermann Minkowski's definition of time as the 'fourth dimension'.
- Ernest Rutherford's detection of a single atom (Rutherford also won the 1908 Nobel Prize in Chemistry for his work in radioactivity).
- Jean-Baptiste Perrin's proof of the existence of molecules (which earned him the 1926 Nobel Prize in Physics).
- Development of Geiger Counter to measure radiation by Hans Geiger and Ernst Rutherford.
- Herman Anschutz-Kampfe's invention of the gyroscope, the compass without a magnetic needle.
- Ikeda Kikunae's discovery of the food additive monosodium glutamate (MSG).
- Sullivan Thomas's invention of tea bags.
- Jacques Brandenberger's invention of cellophane.
- Liquefaction of helium at −268.95°C by Heike Kamerlingh Onnes.
- The discovery of the first large deposit of petroleum in Persia (now Iran), marking the beginning of the Middle East oil boom.
- 'Count von Zeppelin and His Triumphant Airship' (the headline of a major feature article in The New York Times of 5 July 1908).
- Of course, the introduction of Ford's assembly line that rolled out the Model T, the first mass-produced car (which came with Henry Ford's famous promise, 'Any color—so long as it's

black').

The year also had plenty to offer to those who liked their science a little bit spicy. There were stories of captains spooked by 'magnetic clouds' descending on their ships.

The provincial American newspapers, as they still are, were full of reports of 'alien ships' racing at fantastic speeds. (The term 'flying saucer' was coined in June 1947, when an American pilot named Kenneth Arnold saw nine bright discs in the sky near Mount Rainier in Washington. He said the objects were moving like 'a saucer skipping across the water'.)

All this paled into insignificance when the passengers on a steamship in the Gulf of Mexico described in detail the tale of a '200-feet long sea serpent' rising from the sea, which they had seen with their own eyes.

In the history of natural catastrophes, 1908 is remembered for two reasons: the Messina earthquake and the official introduction of Morse code SOS (... --- ...) for the international signal of extreme distress. The most violent earthquake recorded in Europe's history killed 150,000 people in southern Italy and Sicily. The epicentre was Messina, Sicily's second-biggest town.

Had Messina been hit by the Tunguska fireball, the devastation would have been unimaginable. What if the fireball had hit a large city like St Petersburg or London? The sound of the SOS would still be echoing in humanity's ears.

Twenty years later

The first reports of the Tunguska fireball reached the Western world in 1928. The following year, C.J.P. Cave, a British astronomer, noticed the coincidence of the fireball's date and the graphs recorded at six different locations in England on 30 June 1908.

In 1930 another British astronomer, Francis John Welsh Whipple (not the American astronomer Fred Lawrence Whipple who, as we'll see, proposed the 'dirty snowball' model for comets), suggested that the airwaves in England were caused by the great Siberian meteorite, as the

Tunguska fireball was known then. 'How it happened that the fall of the great meteor which produced the waves was not brought to the notice of the scientific world at the time is a mystery,' he said at a meeting of the Royal Meteorological Society. 'There are many marvellous features in the story of the Siberian meteorite, a story without parallel in historical times. It is most remarkable that such an event should occur in our generation, and yet be so nearly ignored.'

Whipple used the six graphs to show that the pressure fronts had been travelling with an average speed of about 1,130 km/h. Pressure waves of similar intensity were recorded in Britain on 27 August 1883 when a volcanic eruption happened on the other side of the world, at Krakatoa on the Indonesian island of Rakata. Whipple also established that the first four almost similar waves were registered within two minutes, and then the final two waves followed. He concluded that there were two kinds of phenomena: the first four waves were caused by a meteorite passing through the atmosphere, and the last two waves referred to the meteorite striking the ground.

After reading Whipple's paper, another astronomer, Spencer Russell, immediately associated the remarkable night glows of 1908 with the Siberian meteorite. 'The entire northern sky on these two nights was of suffused red hue, varying from pink to an intense crimson,' he recalled. 'There was a complete absence of scintillation or flickering, and no tendency for the formation of streamers, or a luminous arch, characteristic of auroral phenomena. Twilight on both of these nights was prolonged to daybreak, and there was no real darkness.'

Though they explained the cause of the phenomena observed in England, neither Whipple nor Spencer made any attempt to explain the nature of the great Siberian meteorite. This task was left to a relatively unknown young Russian scientist.

Chapter 2

The Case of the Missing Meteorite

About 7 p.m., 16 June 1794, Siena, Italy. A vibrant town popular with English tourists. It has a population of nearly 30,000 and a university founded in 1240. Tourists sipping the local Chianti wine in Piazza del Campo, the town's large shell-shaped centre, notice a dark cloud spoiling a picture-perfect blue sky. As the cloud moves southeast, they hear sounds of distant thunder.

A few kilometres away, villagers walking back to their houses in Cosona are bewildered as several stones hiss through the air and land at their feet. Soon afterwards a loud continuous sound, as if from artillery fire, fills the sky. A young woman named Lucrezia Scartelli is curious. She picks up a stone the size of an olive, and immediately drops it as it burns her hand. She hears a thundering sound and sees another large stone falling. She runs away, scared.

Later, another witness named Ferdinand Sguazzini tells other villagers that the stones came so fast that the big ones went right into the ground. As deep as my arms, he says, stretching his arms out. He then shows them a stone: it is the size of a tomato and dark black. He scratches it with his knife and his audience scream with surprise when they find it silvery-white inside.

Stones from the heavens

Ambrogio Sodani, a professor of geology and zoology at the University of Siena at the time, studied the phenomenon and found that the stones fell in an area of 47 square km. He estimated their number to be a few hundred. Their sizes ranged from a small pebble to a 3.5-kilogram rock. He tested them with a magnet and concluded that they were mostly iron. They appeared similar to other stones from the sky that he had seen before.

The famous biologist Lazzaro Spallanzani, who was in Naples in 1788 while Mount Vesuvius was in eruption, suggested that a tornado had carried the Siena stones from Vesuvius, 320 km away, which had again erupted just eighteen hours earlier. But Sodani believed that the stones he had collected were different from the volcanic rocks of Mount Vesuvius. He suggested that they had come from the sky.

Everyone ignored his suggestion except Ernst Florens Friedrich Chladni, a physicist from Wittenberg in Germany (he was a corresponding member of the Academy of Sciences at St Petersburg, and rightly Russia also claims him).

Just two months before the stones fell in Siena, Chladni had published a slim book, *On the Origin of Iron-masses*, in Riga. In this book, he claimed that stones and masses of iron fall from the sky and some of them even create fireballs in the atmosphere. He suggested that these objects originated in 'cosmic space' and might be remnants of planet formation or planetary debris from explosions or collisions.

Chladni's idea that meteorites were extraterrestrial in origin was scientific heresy: it was an attack on the great Newton himself, who believed that apart from the heavenly bodies—stars, planets and comets—all space beyond the Moon was empty (the heavens are devoid of all matter except a very thin, invisible ether, he said in 1704).

The scientists of his time ridiculed Chladni's book. 'By all means, you must read Chladni's infamous book on iron masses,' Alexander von Humboldt wrote to a friend. Georg C. Lichtenberg 'wished Chladni had not written his book.' He felt that Chladni had been 'hit on the head with one of his stones'.

The idea that rocks don't just fall out of the sky was so entrenched that even America's scientifically literate president, Thomas Jefferson, is believed to have commented, 'Gentlemen, I would rather believe that two Yankee professors would lie than believe that stones fall from heaven,' when told that two Yale University professors had reported the fall of meteorites over Weston, Connecticut, in December 1807.

We do not know whether Jefferson's remark is truth or myth. Still, we do know for sure that the French Academy of Science was one of the staunchest critics of Chladni and continued to reject his ideas, even though stones were literally falling in front of bewildered witnesses (at Wold Cottage, England, on 13 December 1795; Evora, Portugal, on 19 February 1796; and Benares, India, on 19 December 1798).

When a spectacular shower of several thousand stones fell near the town of L'Aigle in northern France on 26 April 1803, which was witnessed by many French officials, the Academy hastily dispatched one of its members, physicist Jean Baptiste Biot, to investigate the phenomenon. 'I collected and compared the accounts of the inhabitants: at least I found some of the stones themselves on the spot, and they exhibited to me physical characters which admit of no doubt of the reality of their fall,' Biot wrote in his report. The report finally convinced the scientific establishment that stones do fall from the heavens.

Astronomers now remember Chladni as the founder of meteoritics, the science of meteorites; physicists remember him as the founder of acoustics for his mathematical investigation of sound waves (the patterns formed when a thin plate, covered with sand, is made to vibrate are still called Chladni figures).

A lesson in meteoritics (and investing in metals)

Today, more than two centuries after the publication of Chladni's 'infamous' book, we know that meteorites (from the Greek word *meteoros*, meaning 'high in the air') are chunks of extraterrestrial matter, remnants of geological processes that formed our solar system 4,600 million years ago. When these chunks enter Earth's atmosphere, they shine brightly because of the heat produced by friction with the air. Most pieces are too small—usually the size of a grain of sand, but no larger than a pea—to survive the trip, and are called meteors (or falling stars or shooting stars because they leave momentary streaks of light in the sky).

Very rarely, a large chunk, which flashes like a fireball in the sky, survives its journey through the air to hit the ground. The falling object—a solid piece of stone or iron, often weighing many kilograms—is known as a meteorite. (Until a meteor or a meteorite enters Earth's atmosphere it is known as a meteoroid.) Asteroids (also called planetoids or minor planets), on the other hand, are small bodies orbiting Sun, mostly in between Mars and Jupiter. Most meteorites are pieces of rock and/or metal from asteroids; most meteors are produced when comets disintegrate. Comets are independent masses of ice and dust that orbit Sun.

The fall of a large meteoroid is a rare but spectacular event. The meteoroid enters Earth's atmosphere at a very high speed, ranging from 40,000 to 250,000 km/h. The friction against air not only decelerates it but also raises its temperature. At a height of about 100 km, the meteoroid is so hot that it shines like a fireball. Its outer layer is continuously vaporised and ejected, leaving a trail of dust and smoke. At a height of about 20 km, its speed has been so much reduced—it is about 10,000 km/h—that it no longer glows. It decelerates further to a free-fall speed of somewhere between 320 and 640 km/h until it reaches the ground. Because of high speed and high temperature, most large meteoroids break into several or sometimes thousands of fragments at a height of between 11 and 27 km. The fragments strike Earth in a roughly elliptical pattern a few kilometres long. The 'shower of stones' at Siena in 1794 was this type of fall.

A meteorite's spectacular entry into the atmosphere is accompanied by an equally spectacular sonic boom. Because sound travels quite slowly, at only about 1,200 km/h, it takes from 30 seconds to several minutes after the appearance of the fireball before any sonic boom can be heard. However, many witnesses have claimed that they heard strange noises as a meteorite streaked across the sky.

Known as electrophonic sounds, these range from hissing static to the sound of an express train travelling at high speed. Scientists have not yet validated electrophonic sounds. Still, scientists suspect that light given off by a meteorite must be accompanied by invisible electromagnetic radiation in the form of VLF (very low frequency) radio waves at frequencies from 10 hertz to 30 kilohertz. These waves could reach the observer as soon as the meteorite approached, but you wouldn't hear them. Often, the witness of such sound is located near metal objects. Such objects may act as a transducer, converting inaudible electromagnetic waves into audible sound vibrations.

Meteorites contain various proportions of metals (iron and nickel) and stones (silicates). Thus, meteorites can be divided into three simple categories: irons consist mainly of metals; stones consist of silicates with little metal; and stony irons contain abundant metals and silicates.

The largest known meteorite is still lying where it fell in prehistoric times in Hoba, Namibia. This room-sized meteorite is one metre high and weighs 60 tonnes. It is mostly iron. The most famous sacred meteorite is the black stone of the Ka'bah, which now lies in the Great Mosque in Mecca, Saudi Arabia, towards which Muslims pray five times daily. Islamic tradition has it that the stone came from heaven and was originally hyacinth in colour before it turned black because of humanity's sins.

A large meteorite or asteroid hits the ground with such an enormous force that it shatters into pieces and leaves a big hole—a crater. How would you recognise a meteorite crater if you fell into one? When the meteorite shatters at the moment of impact, the pulverised earth and meteorite fragments are hurled out of the crater and scattered around it. Still, a considerable part falls back into it. This causes craters to display raised and overturned rims. But most of the time this above-surface evidence is erased as Earth's surface is always changing. After thousands of years of weathering and erosion by wind, rain, ice, changes in temperature, gravity and activities of animals and plants, a crater may not look like a crater.

Perhaps the most important characteristic of a crater is the presence of meteorite fragments in the vicinity of it. Only small craters have meteorite fragments. If a meteorite explodes the moment, it strikes the ground, most of it changes into gas. The main feature of craters produced by such meteorites is the complete absence of fragments. Another feature is that these craters have diameters of more than one kilometre. Thus, the absence of meteorite material is not evidence against a meteorite impact.

Even if a meteorite does not leave any fragments, it leaves some evidence of impact. Geologists look for three types of impact evidence:

Impactites. The impact produces so much heat that the rocks melt and splatter into the air. As the drops of melted rock cool they turn into glassy globules, called impactites. They often contain iron-nickel grains, remnants of the meteorites. Depending upon how much glass and other minerals they contain, these globules are sometimes given specific names such as tektites, krystites and suevites.

Shocked quartz. The impact also shatters the rocks, throwing tiny grains of quartz into the air. The shattering is so violent that it leaves patterns on these grains, known as shocked quartz.

Rare elements. Certain elements such as iridium are rare in rocks in Earth's crust. A gigantic impact can scatter these elements all over the world.

Since the late 1950s geologists have confirmed only 160 impact craters. Metal meteorites formed them all, and their diameters range from 10m to 200 km. No craters associated with stony meteorites have been found. The reason perhaps is that stony meteorites disintegrate in the atmosphere. Even if some fragments have survived the journey, they may not have survived the terrestrial weathering and erosion.

The world's first authenticated and best-preserved impact crater is in Arizona. Known simply as the Meteor Crater, its rim-to-rim diameter is 1.2 km and its circumference is nearly 5 km. Its depth below the surrounding plain is about 175 m, with a 45-m-high rim rising above the plain. It was gouged in about 50,000 years ago by a meteorite with the diameter about the width of a football field. The original meteorite, packed with more than 300,000 tonnes of iron and nickel, was travelling with a speed of about 64,000 km/h. It was strong enough to pass through the atmosphere without breaking into pieces.

The Meteor Crater has also found a place in the annals of foolhardy investments. In 1902 Daniel Moreau Barringer, a successful mining engineer from Philadelphia, heard about the crater and the small balls of iron scattered around it. He rejected the prevalent idea that the crater was formed by a volcano and formed the view that it was a meteorite crater. He estimated that an iron-nickel body weighing between 5 and 15 million tonnes lay beneath the surface. In 1903, without ever having seen the crater, he formed the Standard Iron Company and applied for and received from the United States government a 199-year lease (signed by President Theodore Roosevelt himself) on two square miles of land around the crater.

Over the next 26 years, until Barringer's death in 1929, the Standard Iron Company spent more than $600,000 (a considerable fortune in those days) drilling scores of holes. The deepest hole reached 412 m but it produced nothing except tiny samples of meteorite material. The samples contained 93 per cent iron and 7 per cent nickel and traces of other elements, including precious platinum and iridium.

The 'fool's iron' probably gleamed in the eyes of Russian authorities when in 1921 they decided to find meteorite falls which had been recorded in Russia during and after the war years. The new regime formed after the October Revolution of 1917 (the Union of Soviet Socialist Republics was born in 1922) was reeling economically when the news of the iron bonanza of the Meteor Crater reached Moscow. Authorities dreamed about discovering iron worth a fortune that had fallen within Russia. A bright 38-year-old scientist was put in charge of finding these 'treasures from space'.

Our man in Tunguska

Fast forward to 30 June 1958. The face that peers out from the 40-kopeck stamp released today by the USSR is that of a worried, bespectacled, grey-bearded man wearing a fur hat. The stamp commemorates the 50th anniversary of the Tunguska fireball and the scientist who devoted the last two decades of his life to solving the riddle of the mysterious Siberian meteorite. He was the first scientist to visit the Tunguska site.

From 1927, when he led the first expedition to Tunguska, to his death in 1942 in a Nazi prisoner-of-war camp, Leonid Alekseyevich Kulik, a mineralogist and an authority on meteorites, believed that the Tunguska fireball was a giant meteorite. His four expeditions to the site from 1927 to 1939 (the planned fifth expedition was postponed because of the outbreak of the Second World War) failed to find any remains of the meteorite, which he firmly believed had been lying hidden somewhere in the explosion site.

'Where was the meteorite crater with its raised rim that should have been created at the moment of impact?' The question haunted Kulik. The worried face in the photograph on the 1958 stamp seems to ask the question: if it were not a giant meteorite, then what caused the great Siberian explosion? The mystery still eludes scientists. There are many theories, but no definitive answer. But first, the man who made Tunguska famous.

The eldest son of a doctor, Kulik was born on 31 August 1883 in Tartu, Estonia, which later became part of the Soviet Union. After graduating from high school with a gold medal, he studied at the St Petersburg Forestry Institute. In 1904 he was expelled for Bolshevik revolutionary activities. He spent the following year teaching mathematics and physics to adults in a night school. In 1906 he was imprisoned for a short time. After leaving prison, Kulik studied physics and mathematics at Kazan University. He married Lydia Ivanova in 1907, and his first daughter Helen was born in 1910 and his second daughter Irina in 1925.

In 1910 Kulik was again imprisoned for a brief time and remained under police surveillance for two years. In 1912, he went to the Ural Mountains where he worked as a forestry officer. There he met Vladimir I. Vernadsky, a highly respected geochemist and mineralogist, who was leading a group of geologists searching for mineral deposits. This chance meeting was the start of the metamorphosis of a revolutionary and an amateur poet into a pioneering scientist.

Vernadsky was so impressed with young Kulik's quick grasp of mineralogy that he predicted that this 'lover of minerals and nature' would one day become a major scientific researcher. Vernadsky arranged to have Kulik transferred from the forestry department to his expedition, and eventually to the prestigious Mineralogical Museum of the Academy of Sciences at St Petersburg.

The start of the First l War in 1914 put a stop to Kulik's rapid rise as a mineralogist. He was drafted into the Russian Army and fought in East Prussia. After the war, he studied at the military academy and then continued working for the army as a scientist. During the October Revolution, Kulik moved to Tomsk, Siberia's major city, where he joined the Red Army and also taught mineralogy at Tomsk University.

After his discharge from the Red Army in 1920, Kulik returned to his museum post in St Petersburg. With the single-minded intensity that had characterised his life as a revolutionary, army officer and teacher, Kulik now applied himself to the study of meteorites and within a short period established himself as an authority on this relatively new branch of science.

Also working at the museum at the time was Evgeniy L. Krinov, a highly respected mineralogist and an authority on meteorites. Krinov, who is best known for his authoritative book *Giant Meteorites* (published in English in 1966), called Kulik 'a vibrant, cultured man around whom young people flocked'. Krinov also admired Kulik 'as an outspoken individual who was not afraid to voice his opinion when he was convinced, he was right.' An admirable trait in any scientist.

Figures 3. Leonid Alekseyevich Kulik, the first Tunguska researcher.

Figures 4. The Soviet Union stamp honouring Kulik on the fiftieth anniversary of the Tunguska event in 1958.

The meteorite hunter

In 1921 the Soviet Academy of Sciences approved the country's first special meteorite expedition, which was charged with the task of locating and examining meteorites fallen in inhabited regions of Russia. One of the expedition's tasks was to gather information from local populations and talk to eyewitnesses. The expedition left Petrograd (as St Petersburg was known after the First World War; the old name sounded too German for contemporary Russians) on 5 September 1921 under the leadership of Kulik.

At this time Kulik was not aware of the Tunguska meteorite. He first heard about it at the railway station as the small expedition party set off on the Trans-Siberian Express. D.O. Sviatsky, the editor of the magazine *Mirovedeniye*, ran up to the train and gave Kulik a page torn from the 1910 calendar published by Otto Kirchner of St Petersburg. On the back of this page was the following note:

> About 8 a.m. in the middle of June 1908 a huge meteorite is said to have fallen in Tomsk, several sagenes [1 sagene = 2.314m] from the railway line near Filimonovo junction and less than 11 verst [1 verst = 1.067 km] from Kansk. Its fall was accompanied by a frightful roar and a deafening crash, which was heard more than 40 versts away. The passengers of a train approaching the junction at the time were struck by the unusual noise. The driver stopped the train and the passengers poured out to examine the fallen object, but they were unable to study the meteorite closely because it was red-hot. Later, when it had cooled, various men from the junction and engineers from the railway examined it, and probably dug around it. According to these people, the meteorite was almost entirely buried in the ground, and only the top of it protruded. It was a stone block, whitish in colour, and as much as six cubic sagenes in size.

Kulik was fascinated by the story, so fascinated that he immediately decided to investigate it further. Over the years, his fascination would become an obsession that would consume the rest of his professional life.

The Academy of Sciences had provided Kulik's team with a carriage on the Trans-Siberian Express. They travelled across the Urals into Siberia, then made stops in Omsk, Tomsk and Krasnoyarsk, and finally arrived at Kansk. At Kansk, Kulik searched through Siberian newspapers published during the summer of 1908 for reports of a meteorite fall. He soon discovered that the calendar note was the beginning of an article published on 12 July 1908 in the newspaper *Sibirskaya Zhizn* from the Tomsk region. The report turned out to be wrong in almost every detail, except about the train stopping near Kansk.

As he sifted through newspapers, he found many reports of a huge meteorite fall on the morning of 30 June 1908. He prepared a questionnaire and published it in local newspapers and distributed 2,500 copies among the locals. As a result, he collected breath-taking personal accounts, which were vivid and rich in details, from several dozen eyewitnesses who could still remember the event. From the information collected, Kulik painstakingly painted a picture of the meteorite, which he called the 'Filimonovo meteorite' (the term 'Tunguska meteorite' was not used until many years after the first Tunguska expedition in 1927).

Though he now firmly believed that between 5.00 and 8.00 a.m. on 30 June 1908 a giant meteorite flew in the general direction of south to north and fell probably in the basin of the Vanavara River, a tributary of the Podkamennaya Tunguska River, he was unable to embark upon a search. The expedition had run out of funds, and the authorities needed his train carriage which had been lent for the expedition.

On return to Petrograd, Kulik submitted his report to the Academy of Sciences in which he suggested that the Siberian meteorite was a rare event in human history and must be investigated. His 'Account of the Meteorite Expedition' was published in the *Journal of the Soviet Academy of Sciences*. The Academy members, however, were sceptical of the claim. But Kulik was not alone in his quest.

A.V. Voznesensky, who was Director of the Irkutsk Magnetic and Meteorological Observatory in 1908, published a report in *Mirovedeniye* in August 1925 in which he claimed that the seismic and airwaves recorded by his observatory on 30 June 1908 were both caused by the fall of a giant meteorite. He suggested that the airwaves were caused by 'the explosion of the meteorite at a height of about 30 km above the surface of Earth.' In his report, Voznesensky also rightly pointed out that his observatory's seismograph registration of the fall of the meteorite was the first in the history of science.

Voznesensky also suggested that the investigator of the spot where the Siberian meteorite fell would find something very similar to the meteorite crater of Arizona. 'The Indians of Arizona still preserve the legend that their ancestors saw a fiery chariot fall from the sky and penetrate the ground at the spot where the crater is; the present-day Evenk people have a similar legend about a new fiery stone,' he said. He concluded his report with a tantalising idea. In essence, the search for the meteorite could prove a very profitable enterprise, mainly if this meteorite turned out to belong to the iron class.

In the same issue of *Mirovedeniye*, S.V. Obruchev, a geologist, wrote about his studies in the summer of 1924 in the Tunguska region. He also described stories of a huge calamity which had been told to him by the local indigenous inhabitants, the Tungus people (later named Evenk by the Soviets; they are probably the oldest surviving native Siberians). Obruchev speculated that a giant meteorite caused the calamity. 'In the eyes of the Tungus people, the meteorite is apparently sacred, and they carefully conceal the place where it fell,' he said. However, he learned that there was a 'flattened forest' three- or four-days northeast of Vanavara.

In 1926, I.M. Suslov, an ethnographer, visited the Tunguska region. In his report 'In Search of the Great Meteorite of 1908' in *Mirovedeniye* (March 1927), he described some of the 60 eyewitness accounts of the explosion he had collected. In these accounts, he said, such expressions were heard as 'the forest was crushed,' 'the grain stores were destroyed,' 'the reindeer were annihilated,' 'people were injured,' 'the taiga was flattened,' and so forth.

Suslov also visited the tent of Ilya Potapovich Petrov, the very same Evenk whom Obruchev had questioned in 1924. Ilya Potapovich, who would work as a guide on Kulik's 1927 expedition, agreed to Suslov's request to draw a map of the area of the catastrophe. 'Ilya Potapovich drew the map with coloured pencils, and a group of Evenk people made corrections,' Suslov said.

The articles of Voznesensky, Obruchev and Suslov convinced the Soviet Academy that an event of major importance had occurred, and investigations should be continued. Kulik's mentor Vernadsky also supported his request. 'The expedition proposed by Kulik may turn out to have a very great scientific significance, and its results may repay a hundredfold the time and money spent on it,' he wrote to the Academy. The Academy approved the first Tunguska expedition in 1926.

The first Tunguska expedition

In early February 1927, Kulik left Leningrad (as St Petersburg was again renamed in 1924) with one assistant named Gyulikh. Travelling by the Trans-Siberian Express, on 12 February he reached the remote Siberian station of Taishet (also spelled Tayshet), some 900 km south of the Tunguska explosion site. After buying food and supplies and other equipment, Kulik and his assistant left Taishet by horse-drawn sledges. Battling frequent snowstorms and bitter temperatures, it took them five days to reach the small village of Kezhma, about 215 km south of the explosion site, where they replenished their food and supplies and left with three carts on 22 March.

Perhaps Kulik had read Nikolai Gogol's novel *Dead Souls* that evokes the misery of Siberia, 'I see you: a country of dinginess, and bleakness and dispersal,' and knew what lay ahead.

No one has better described inhospitable Siberia than the Russian writer Anton Chekhov. He was 30 years old when in 1890 he made an incredible trip, mainly by horse carriage and riverboat, across Siberia to the island of Sakhalin, a penal colony in tsarist Russia. 'Why is it so cold in this Siberia of yours?' With this question of his coach driver, Chekhov's journal of his expedition, *A Journey to Sakhalin*, begins. 'Because that's the way God wants it,' replies the driver. He travelled in relatively more inhabited areas of Siberia, yet Chekhov complained: 'Siberian highways have their little scurvy stations ... They pop up every 20 or 25 miles. You drive at night, on and on, until you feel giddy and ill, but you keep on going, and if you venture to ask the driver, how many miles it is to the next station, he invariably says, "Not less than twelve".'

Kulik did not have the luxury of a coach or the opportunity to complain to someone. He and his assistant were now travelling through the most rugged taiga, split with creeks, gullies, bogs and swamps and steep hillsides. They had to take many detours to ford rivers because it was too dangerous to cross flimsy suspension bridges. They were now in a 'vast and sinister' primeval forest in which 'the weak and imprudent often perish', as described by the Russian writer Yuri Semyonov in his book *The Conquest of Siberia* (1944). Nevertheless, three days later they arrived in Vanavara, the most northerly outpost of civilisation, a tiny trading station of only a few houses and stores, situated on the high right bank of the Podkamennaya Tunguska River.

Kulik had been carrying a letter from Suslov to the local Soviet political officer with a request to put Kulik in touch with the Evenk Ilya Potapovich. To Kulik's dismay, Ilya Potapovich flatly refused to guide him to the 'thunder god's home,' the forbidden and sacred land. To the Evenk people, the fiery body was a visitation from Ogdy, their god of thunder, who had cursed the area by smashing trees and killing the animals. No one dared approach the site for fear of being cursed by Ogdy.

In an article in *Astronomy & Astrophysics* (February 2008), the Russian nuclear physicist Z.K. Silagadze describes an Evenk myth related in 1977 by G. Gordon and K. Monakan on their (now defunct) website, 'The Tunguska Project'. The tale highlights the spiritual significance of the Tunguska explosion to the Evenk people. The story starts with the battle between with the battle between two Tunguska Evenk clans:

'Over the years, their feud escalated, both clans using their powerful shamans to curse to the other, with evil spirits, misfortune and disease. The hostility between them grew until one shaman called upon the Agdy to destroy the hated enemy forever. These fearsome iron birds fly above the earth in huge clouds, flapping their terrible wings to cause thunder, flashing lightning from their fiery eyes. On that sunny morning in June, the sky became black as a never-ending legion of the fearsome birds swooped low over the unfortunate Shanyagir clan. Their devastating blasts of fire blew the Shanyagir's tents up into the air over the treetops. The clan's belongings were destroyed, 250 of their reindeer vanished without a trace, the ancient forest was flattened in every direction, and those who still could fled in panic. To this day, the Evenk believe that only the Agdy can live in the area where the explosion took place. Only a few will risk visiting. And none will live there.'

Kulik's dogged determination eventually won over Ilya Potapovich's reluctance when he offered him two bags of flour, several rolls of cloth and building materials for the roof and floor of his house.

Kulik was keen to continue the journey immediately, and the day after arriving at Vanavara, he set out with his assistant and new guide. Their horses were tired after the long journey, and being overloaded, they could not make their way through the snow-covered forest. They were forced to return to Vanavara and await better weather.

On 8 April Kulik's party set off again. This time they were better prepared, and their packhorses were loaded with enough food to last about a month. They travelled along the Podkamennaya Tunguska River for about 30 km downstream until they reached the Chamba River. They then followed this river for about 10 km and by nightfall reached the hut of the Evenk Okhchen, who agreed to join them as a second guide.

The next morning, they reloaded all their supplies onto Okhchen's flock of ten reindeer and set off over a reindeer track along the Chamba River. Two days later the track came to an end. The five days' arduous journey had taken its toll. Exhausted and sick with scurvy and various infections from months of poor food, Kulik was still determined to go on. They hacked their way with axes through the virgin taiga.

On 13 April the expedition crossed the Makrita River, where they found the beginning of a mass of fallen trees, uprooted as by an explosion.

In the distance could be seen the twin-peaked mountain called Shakrama by the Evenk people. On 15 April Kulik climbed the mountain and saw the explosion site stretching to the horizon before him. 'This is where the thunder and lightning fell,' pointed out Ilya Potapovich, 'and burned down my relative Onkoul's grain store'. (Kulik's third expedition indeed discovered the remains of the grain store.)

Kulik saw an oval plateau 70 km wide where the forest had been flattened. Trees were not uprooted: instead, they were stripped of their branches, snapped off and scattered like matchsticks pointing away from the direction of the blast. 'The results of even a cursory examination exceeded all the tales of the eyewitnesses and my wildest expectation,' Kulik wrote in his diary. 'One has an uncanny feeling when one sees 50- to 75-cm thick giant trees snapped across like twigs, and their tops hurled many metres away to the south.'

Kulik wanted to explore the centre of the blast area, which he assumed lay beyond the distant snow-covered ridges to the north where the forest had been completely destroyed, but his Evenk guides were extremely superstitious and refused to walk through the taiga burned by their god Ogdy. He had no choice but to return to Vanavara.

Figures 5. Charred and fallen trees near the blast site as seen by Kulik. (Photo by N.A. Strukov, Moscow, 1928.)

Kulik was exhausted but determined to find the fall point. Back in Vanavara on 22 April, he hired Russian peasants from Kezhma village and planned a new route to the explosion site. The expedition left Vanavara on 30 April. After three days' journey by sledge they again reached the Chamba River. This time Kulik decided to build rafts and navigate first up the Chamba and then along the Khushmo River, which were flooded with the thawing snow. Ten days later the expedition reached the mouth of the Churgima River, a tributary of the Khushmo. On 20 May they finally arrived at the edge of the devastated taiga. Kulik decided to camp there.

The next day as he followed the direction of the fallen trees for a few kilometres, he reached a marshy basin between 5 and 7 km in diameter and surrounded by low-lying hills. In their legends, the Evenk people referred to the area as the South Swamp, but to Kulik, it resembled a gigantic cauldron, and he named it the Great Cauldron. Here the devastation was greater than what he had seen from Mount Shakrama. Kulik decided to transfer his camp there. Over the next few days, he walked around the hills, climbed them and measured the direction of fallen trees. Kulik was now convinced that he had found the epicentre of the fall. 'There could be no doubt,' he wrote later. 'I had circled the centre of the fall. With a fiery stream of hot gases and cold solid bodies, the meteorite had struck the cauldron, with its hills, tundra and swamp.'

Figure 6. The forest of 'telegraph poles' as seen by Kulik. (Photo by I. M. Suslov, Moscow, 1928.)

Everywhere, for a distance of more than 30 km from the centre, was like a forest of 'telegraph poles', dead trees still standing, but their twigs and branches blown away. 'The taiga has been practically destroyed by being completely flattened,' he recorded in his diary. 'The trees lie in rows on the ground, without branches or bark, in the direction opposite to the centre of the fall. This peculiar "fan" pattern of fallen trees can be seen very well from some of the heights that form the peripheral ring of trees.'

There was another remarkable feature: within the central blasted area was a ring of upright trees, completely stripped of foliage. The fact that they had remained upright while all trees outside the ring had been flattened, Kulik thought, marked some node or region of rest where airwaves cancelled each other.

There was also evidence of fire; some of the trees were charred and lifeless. But this evidence of burning was unusual: in a forest fire, trees are usually burnt on the lower part of their trunks, but these trees had been burnt uniformly and continuously. Kulik believed that a great rush of hot air produced by the change of kinetic energy into heat energy when the meteorite crashed into Earth blew the trees down and scorched them.

In some areas, Kulik also found forest growth about 20 years old. 'From our observation point no sign of forest can be seen, for everything has been devastated and burned,' he wrote in his diary, 'and around the edges of this dead area the young 20-year-old forest growth has moved forward furiously, seeking sunshine and life.'

Kulik also noticed giant circular ridges, like waves in water, which he believed were formed when the solid ground heaved outwards under the impact of the meteorite. The whole scene was like a giant picture of what happens when a brick from a wall falls into a puddle of mud. Kulik had expected to find the evidence of a giant meteorite in the central part of the basin but found that the area was dotted with dozens of holes 'exactly like lunar craters'. These funnel-shaped holes ranged from 10 to 50 m across and up to 4 m deep. Their edges were mostly steep, the bottoms flat and swampy.

'I cannot say how deeply the meteorites had gone into the tundra and the rocks,' he wrote in his account of the expedition, *Beyond the Tunguska Meteorite*. 'It was impossible for me to go right around the whole area ... or do any digging. We had food left for only three or four days, our road was a long one, and our one thought now was to get back safely. It was a flight in the full sense of the word.'

He arrived at Vanavara at the end of June after nine days of travelling. There followed another three weeks on a raft on the Podkamennaya Tunguska River to the town of Yenisei, and then a comfortable journey by steamship to Krasnoyarsk and by train to Leningrad. Kulik was already dreaming about his next expedition.

In his report to the Academy of Sciences he wrote: 'This picture [of shallow holes] corresponds exactly to the theoretical conditions of fall of a swarm of large meteorite fragments, the larger specimens of which exceeded 130 tons. In all probability, these fragments were of iron meteorites ... giant craters such as the Arizona crater are strewn with fragments of iron meteorite.' He concluded his report by pointing out: 'Since this fall occurred on the territory of the Soviet Union we are duty bound to study it.'

On 13 March 1928, the 'duty bound' Academy approved the second expedition to continue the study of the Tunguska meteorite. The Academy seemed to agree with Kulik's appeal in his report. In essence, the significance of the Tunguska fall 'will be fully appreciated only in history, and it is necessary to record all the remaining traces of this fall for posterity'. However, the Academy granted limited funding that allowed only for mapping the blast area and magnetic survey of the holes. Kulik was also expected to recover meteorite fragments for the Mineralogical Museum.

Kulik's first expedition also attracted the attention of the Western press.

In a detailed scientific article entitled 'The Great Siberian Meteorite: An Account of the Most Remarkable Astronomical Event of the Twentieth Century,' in *Scientific American* (July 1928), Chas P. Olivier of the International Astronomical Association wrote: 'Fortunately for humanity, this meteoric fall happened in a region where there were no inhabitants ... but if such a thing can happen in Siberia there is no known reason why the same could not happen in the United States.'

The Literary Digest of 30 June 1928 also warned: 'Had chance directed this enormous visitor from space to the site of a city or a thickly settled country the world would have experienced an unparalleled disaster; one, we must not forget, which may yet happen should another such meteorite ever arrives.'

Nearly a century later, these doomsday warnings remain 'current'—swap the word 'meteorite' with 'asteroid'.

The second Tunguska expedition

In April 1928 Kulik left Leningrad on his second expedition to the Tunguska region. He was accompanied by his assistant V. Sytin, a hunter and zoologist. Kulik did not have any experts from other disciplines of science to study the explosion site. At Vanavara he was joined by a cinematographer, N. Strukov, from Moscow's Sovkino Studio.

Spring floods delayed the expedition's progress and the group, including five workers, reached the explosion site in late June. Kulik set up his camp and started his investigations. He surveyed an area of 100 square km and marked 150 craters with wooden stakes. He also tried to dig into two of the craters, but the water and the boggy soil made digging impossible. He did not have any pumps to drain water from the holes. His primitive magnetic instruments failed to detect any metal pieces 'brighter than the blade of a knife and resembling in colour a silver coin' which some Evenks had reported finding in the devastated forest.

After a few weeks, Strukov left, accompanied by three workers. Kulik, Sytin and two workers remained. Kulik continued to collect samples of peat and other plant materials for microscopic examination back in Leningrad. By the beginning of August, Sytin and two workers were showing signs of vitamin deficiency. Kulik was also running out of funds. His prospects of finding the meteorite were dim. Kulik found himself in a dilemma. He knew if he went back without any results, his opponents in the Academy would deny him any funds for future expeditions. He worked out a strategy: he would stay behind but would send two workers to Vanavara and Sytin to Moscow to convince the Academy to approve more funding.

Sytin's arrival in Moscow coincided with the dramatic rescue of the crew of the airship *Italia*, which had crashed near the North Pole, by the Soviet icebreaker *Krasin*. Moscow newspapers had found another sensational story of adventure when they got hold of Sytin's story of a Russian scientist who was risking his life in the wilds of Siberia to find a mysterious visitor from space. And his exhausted and sick assistant had come to Moscow to seek funds.

The Academy of Sciences buckled under public opinion. Funding was immediately approved for a rescue expedition and to continue the investigation. Sytin arrived back at the explosion site accompanied by ethnographer Suslov and a group of journalists. Suslov had never visited the place he had written about in *Mirovedeniye*. Kulik was so pleased to see Suslov that he named the largest hole the 'Suslov crater'.

Kulik quickly put everyone to work helping with magnetic measurements of various holes. He started with the 50-metre Suslov crater. With the journalists watching, he hoped to make a discovery but was disappointed not to find any trace of metal in the hole.

Figure 7. A 2004 photograph of the Suslov crater. (Photo by V. Romeiko, Moscow, 2004.)

At the end of October, the expedition returned to Vanavara. When he arrived in Leningrad at the end of November, Kulik was a national hero. He didn't need spinmeisters to spin the Tunguska meteorite story. Strukov's short film of his journey, *In Search of the Tunguska Meteorite*, was also a big help in advancing Kulik's cause.

British and American newspapers widely reported Kulik's second expedition.

The London *Times* of 26 November 1928 noted that Professor Kulik 'has reached Krasnoyarsk ... having tolerably recovered from the hardships of his journey'.

The New York Times reported on 2 December: 'Professor Leonid Kulik, Russian geologist, is now reported to be on his way to Leningrad from the depths of North-eastern Siberia, where a relief expedition found him two months ago after he had been given for dead.' A phrase in the report's long headline also hinted at the treasures buried in meteorite craters:

'METEORIC IRON DEPOSITS ESTIMATED AT $1,000,000'.

The Literary Digest of 16 March 1929 published an account of an interview with Kulik and Sytin, which concluded with the following comment: 'The value of the metals in the Siberian find is estimated by Mr Sitin [sic] as between one hundred million and two hundred million dollars, chiefly for the iron and platinum ... the chief object of further investigation of the site ... is not the recovery of any valuable materials that may exist, but the obtaining of further scientific information about an event almost unique in the recorded history.' It seems that the stories of Barringer's holes in the Meteor Crater were still uppermost in people's minds.

Kulik was an excellent writer and speaker and made meteorites popular among the Soviet population. He made his Moscow audience shiver when, in a lecture, accompanied by Strukov's 'moving pictures of the appalling desolation,' he remarked: 'Thus, had this meteorite fallen in central Belgium, there would have been no living creature left in the whole country; on London, none left alive south of Manchester or east of Bristol. Had it fallen on New York, Philadelphia might have escaped with only its windows shattered, and New Haven and Boston escaped too. But all life in the central area of the meteor's impact would have been blotted out instantaneously.'

There were critics as well. A few geologists continued to voice their doubts about a meteoritic origin of the Tunguska blast. They explained Kulik's 'craters' as the result of permafrost and pointed out that similar holes are often found in other parts of Siberia. The south facing slopes of northern rivers in Siberia, which includes Podkamennaya Tunguska River, are most vulnerable to permafrost. Nevertheless, the Academy of Sciences approved the third expedition under the leadership of Kulik.

The third Tunguska expedition

The third expedition left Leningrad in February 1929 and returned in October 1930. It was a much larger undertaking which included many scientists. Among them was young Krinov (who lost a toe to frostbite on the expedition and who would become the Chairman of the Committee on Meteorites of the Academy of Sciences). The expedition was also well equipped. A horse train of 50 carts carried drilling machines, water pumps, geological, meteorological and surveying instruments, cameras, field tools and other supplies from Taishet, the last station on the Trans-Siberian railways, to the Tunguska site.

Kulik was convinced that the Suslov crater and a chain of craters around it were formed by the fall of separate large meteorite pieces. He decided to dig the Suslov crater first; his team drilling a 4-m hole in it. The digging of this hole took Kulik's men one month's hard work, but they did not find any impact features. The walls showed only undisturbed material. However, they found a rotten tree stump at the bottom of the hole, suggesting that the crater was not caused by the impact of a piece of the meteorite after it broke on hitting Earth. 'Indeed, it was impossible to imagine that a tree stump could have been presented in a natural position so near the centre of the hole formed by the fall of a meteorite,' Krinov noted.

Krinov, who had been making an independent survey of the area, concluded that the epicentre of the fall did not lie on the northern border of the basin, the location of the Suslov crater, as believed by Kulik. He suggested that the exact location of the epicentre was South Swamp, an area a few kilometres south of the Suslov crater. Krinov's suggestion made Kulik angry, and he excluded him from further work on the expedition.

Kulik's belief in his Suslov crater hypothesis was unshakeable. He directed his men to set the drilling rig at the edge of the crater. They continued digging during the frigid winter months and drilled one hole 34 m deep and 4 m wide in the solid frozen ground. Still, they did not find any meteorite material.

Kulik's spirits soared when a worker found a piece of glass in the hole. Kulik concluded that it was an impactite, a rock fused into glass by the impact of the meteorite. But the glass did not turn out to be the smoking gun Kulik was looking for. It was a piece of a bottle shattered during a fire which broke out in one of the huts on the first night of the expedition.

'Unfortunately, this fragment was mentioned in several articles by Kulik as a find of silica glass, and even today it misleads researchers,' Krinov said.

Kulik's team drilled two more holes before they gave up drilling on 1 March 1930. He finally concluded that the Suslov crater was not a meteorite crater. Kulik learned the hard way what is now known to every student of meteoritics: finding a meteorite crater may not be as simple as finding a hole in the ground.

Six months later Kulik returned to Leningrad 'with grey hair and ruined health'. J.G. Crowther, a British science journalist who interviewed Kulik after the third expedition, wrote in *Scientific American* (May 1931): 'Professor Kulik's expeditions have left a mark on him. He is a tall, wiry, bronzed man of Scots figure, lean and a little tired. Perhaps a rest will soon entirely refresh him.'

Crowther's hope turned out to be a prophesy. Kulik was not to return to Tunguska for another seven years, but he continued working on the project. He also apologised to Krinov and requested him to continue working with him. Like Krinov, he now believed that the South Swamp was indeed the blast's centre.

Figure 8. Map of the Tunguska site produced from the visual survey during Kulik's third expedition in 1929-30. (Key: 1—devastated forest; 2—meteorite impact site; 3—survey points; 4—track; 5—road to Strelka trading station; 6—limit of scorched area; 7—limit of forest devastation; 8—limit of the effect of the explosive wave.) (Courtesy Marek Zbik.)

Figure 9. Radial pattern of tree damage. (Courtesy Marek Zbik.)

Solving the riddle

In 1938 an aerial photographic survey of the Tunguska region was undertaken. Although the survey was incomplete, it showed the unique radial nature of the fallen trees. It also showed that the South Swamp was indeed the centre of some great catastrophe.

In July 1939 a 'refreshed' Kulik again returned to Tunguska to examine the South Swamp. His team started boring in several places in the swamp. The drilling showed numerous channels on the surface of the swamp. Kulik interpreted them as underwater craters. This idea was later rejected by geologists who said that this was a natural feature of the swamps in the area. However, the Academy of Sciences congratulated Kulik for his 'great persistence and enthusiasm that led to the recent concrete advance in our knowledge of the subject' and approved the fifth expedition in 1940 to conduct a magnetic survey of the South Swamp. But the expedition did not take place because of the outbreak of the Second World War.

On 5 July 1941, the day the Nazis invaded Russia, Kulik joined the Moscow People's Militia, a volunteer military unit. In October, while taking part in a battle on the front line, he was wounded in the leg and captured by the advancing German army. He was held as a prisoner of war in the town of Spas-Demensk, about 300 km southwest of Moscow. He worked as a nurse in a prisoner-of-war hospital, where he contracted typhus and died on 14 April 1942. He was 58 years old. He was buried in the local cemetery. In 1960 the Academy of Sciences built a simple memorial on his grave. The plain gravestone is still there; it's simply marked: 'Kulik, Leonid Alekseyevich 1883-1942.'

Figure 10. Kulik's grave in Spas-Demensk, Russia. (Photo by Andrei Ol'khovatov, Moscow.)

'Kulik,' a 58-m diameter crater on the far side of the Moon, asteroid '2794 Kulik' and a Vanavara street named after him perpetuate the memory of the first Tunguska researcher and the founder of meteoritics in Russia.

The first expedition after the war was in 1958. In 1963 investigations gained new vigour under the leadership of Nikolai Vasilyev (1930-2001) of the Academy of Sciences, who coordinated the scientific research of 29 investigations. It wasn't until 1989 that foreign scientists were officially invited to join the Russian investigations. Now the Russian government has set aside 4,000 square km of the Tunguska region as a national reserve. But the Tunguska explosion site remains inaccessible. The tiny trading post at Vanavara has now grown into a small town of nearly 4,000 people, which also has an airport. From Vanavara the explosion site is about 70 km, but it can be reached only by helicopter or by hiking.

Why did Kulik fail to find any meteorite fragments or impact crater in the South Swamp or the Great Cauldron? According to Krinov, careful investigation of the cauldron 'does not give any grounds for concluding that this cauldron is the place where the meteorite fell'. But four observations point to the fact that the cauldron is the site of the explosion:

- The absence of other places in the Tunguska area which might attract attention as the possible place of the fall.
- The Evenk people's designation of the cauldron as the place of the fall.
- The cauldron is the epicentre of the seismic wave.
- The radial forest devastation around the cauldron.

'There is only one possible explanation that removes the contradiction, that is, that the meteorite did not explode on the surface of the ground, but in the air at a certain height above the cauldron,' Krinov concluded.

Krinov's explanation did not solve the riddle of the Great Siberian explosion. The controversy about the Tunguska fireball continues to this day, and there is no shortage of attempts to explain the cataclysmic explosion. Igor Zotkin, a Russian expert on meteorites, once remarked: 'I doubt if there is any recent discovery that has not been called on to explain the Tunguska enigma.'

Today, scientists' line-up of suspects includes a comet, a mini black hole, an asteroid, a rock of antimatter, a mirror matter or dark matter, and a methane gas blast from below. In X-files, we have an alien spacecraft, a laser beam fired by ETs in an attempt to communicate with lonely little earthlings, and an experiment on a 'death ray' which got out of hand.

Chapter 3
The Tale of a Fiery Comet

In the 1880s H.H. Warner, a wealthy American renowned as the 'patent medicine king,' awarded a cash prize of $200 to an American or Canadian discoverer of a comet. The prize motivated Edward Emerson Barnard, a young amateur astronomer with virtually no formal education, to discover eight comets within six years. He won enough money to build himself a house. 'This fact goes to prove the great error,' he said, 'of those scientific men who figure out that a comet is but a flimsy affair after all—for here was a house, albeit a small one, built entirely out of them. True, it took several good-sized comets to do it, but it was done nevertheless.'

Barnard's discoveries also earned him a scholarship to Vanderbilt University. In 1887 the 30-year-old graduate joined the Lick Observatory of the University of California at Santa Cruz. In his resumé he listed all his discoveries—10 comets and 23 nebulae—as well as his good habits, 'I am perfectly temperate, neither smoke, chew, nor use intoxicating drinks'.

Barnard, 'the man who was never known to sleep', was well known as an inexhaustible observer of the heavens when he became the subject of an elaborate hoax.

On 8 March 1891, when he opened his copy of the *San Francisco Examiner*, Barnard was astonished and horrified to read a story describing his 'invention of a machine for scanning the skies and catching wandering comets on the photographic plate':

Discovers comets all by itself
The meteor gets in range, 'Electricity does the rest.'
A Wonderful Scientific Invention that will do away with the
Astronomer's Weary Hours of Searching—The Idea Founded
on the Spectrum of the Comet's Light—It's Just Like

Gunning for Wandering Stars with a Telescope.

The long, breezy headline was followed by an equally long and playful story that filled two whole columns of the newspaper and included three detailed illustrations: a view of the complete comet-seeker, a diagram of an objective prism and an electrical circuit. The story also included several quotes from Barnard, whom the story described as 'the renowned young astronomer'. 'When the comet is caught, as in a trap ... An alarm-bell rings in my bedroom down at the cottage,' Barnard was quoted as saying. 'Of course, the signal quickly summons me to the roof [of the observatory] ... A single glance should suffice to reveal the position of the new comet.'

Barnard immediately sent angry letters of denial to all San Francisco newspapers, but they all ignored them. Somehow the hoaxer had convinced the newspapers to ignore Barnard if he 'disowns his invention'. For two years, until the *Examiner* apologised in an editorial after he discovered the fifth moon of Jupiter, Barnard continued to receive letters from all over the world. Even the famous astronomer Lewis Swift wrote that he had read the article 'regarding your invention to search for comets while asleep or using the 12-inch or playing poker ... although the article appears somewhat fishy, I am inclined to think it is still another of the marvellous inventions of the 19th century'.

Barnard never discovered the identity of the perpetrator of the hoax; however, he suspected one of his colleagues. Barnard died in 1923 leaving an astonishing legacy of observations—of planets, satellites, comets, double stars, bright and dark nebulae and globular clusters—that make him one of the greatest observers of all time.

Carolyn's comet

Barnard was 20 years old when he discovered his first comet in 1881; Carolyn Shoemaker was 54 when she found her first comet in 1983. Though she doesn't collect a 'cash prize of $200' for each of her discoveries, she has now 32 comets to her name, more than any other astronomer, living or dead (Jean-Louis Pons, a 19th-century amateur astronomer, discovered 37 comets but only 26 bear his name).

After spending 25 years as 'homemaker and mother' to her three children, Carolyn joined her better-known husband Eugene in 1980 in his search for comets and asteroids. Eugene, a geologist who is considered the father of planetary impact geology and has 29 comets to his name, was killed in a car accident in 1997 while hunting for impact craters in outback Australia.

Carolyn shares her most famous comet discovery with her husband and David Levy, an amateur astronomer who has 21 comets to his name, 13 with Eugene and Carolyn Shoemaker. On 25 March 1993, Carolyn was as usual scanning films of the sky taken the previous night by Eugene and David from the 18-inch Schmidt telescope at Palomar Observatory in California. Unlike the observatory's main 200-inch Hale telescope which can see deep into space, the old Schmidt is an ideal telescope for surveying a wide section of sky.

There is still no 'automatic comet-seeker' to help Carolyn. She spends long hours in a dark room scrutinising pairs of photographs of the same area of sky with a stereomicroscope. The pairs of photographs are taken 45 minutes apart. As the comet will move across the sky in 45 minutes, it will be in a different position on the second film. The stereomicroscope allows Carolyn to see the two films simultaneously—one by the left eye and the other by the right eye. When viewed this way, a comet would appear to 'float' above the flat surface of the fixed stars. But it is not as simple as it sounds: a speck of dust, a satellite or a spark of light in the telescope can make comet-hunting a painstakingly slow sport (Levy considers it a 'competitive sport').

As she was slowly moving the film through the stereomicroscope, she saw a blob that looked like 'a squashed comet'. When Eugene and David looked at it, they were surprised. It was the strangest thing any of them had ever seen during all their collective years of comet hunting. They had just discovered the comet of the century.

Following the tradition, the comet was named after its discoverers— Periodic Comet Shoemaker-Levy 9 1993e (it was the ninth comet discovered jointly by the Shoemakers and Levy which travelled around the Sun in a short-period orbit; and the fifth comet discovered in 1993, hence 'e,' the fifth letter).

When other astronomers trained their telescopes on this curious find, they noticed that Shoemaker-Levy 9, or S-L 9 for short, was a string of 21 comet fragments stretched out in a trail nearly 200,000 km long. S-L 9 had two unusual features. First, it had 21 separate nuclei like pearls on a necklace; no comet observed had broken into as many pieces. Many comets break up when they come close to Sun, but they usually break into two or three pieces. Second, comets usually orbit Sun; S-L 9 was orbiting Jupiter.

Astronomers also calculated that S-L 9 would inevitably crash into Jupiter in July 1994. Carolyn Shoemaker is not used to losing her comets. 'If I am going to lose a comet, then I want it to go out with fireworks,' she hoped. That's just what happened: S-L 9 died in an extraordinary cosmic firework, which Carolyn watched from a safe distance of a billion kilometres.

By Jove, yes, it was the show of shows

In July 1994 the Hubble Space Telescope, orbiting 600 km above Earth, and hundreds of thousands of telescopes around the world were aimed at Jupiter to watch the celestial drama of the century. Never in history had anyone witnessed the cataclysmic collision of two worlds, a comet crashing into a planet.

A few days before the crash, the observers noticed a trail of 21 comet pieces stretching for a length of more than 5 million km, more than 12 times the distance between Earth and the Moon. On 16 July 1994, the first piece with a width of one kilometre—the puniest in the line-up—slammed into Jupiter. Over the next six days, 20 more pieces hit the giant planet.

When plunged into Jupiter's visible clouds, each piece of S-L 9 was travelling with a speed of 200,000 km/h. As the piece entered the denser atmosphere it flattened like a pancake and then disintegrated, dumping energy equivalent to a blast of more than a million megatons (explosive energy is measured in kilotons or megatons; one kiloton equals 1,000 tonnes and one megaton one million tonnes of high explosive TNT). Taken all together, the 21 fragments released about 40 million megatons of explosive energy. Compare this energy with the Hiroshima atomic bomb: the mere 15-kiloton blast killed 140,000 people, injured hundreds of thousands and destroyed 70,000 buildings. The largest hydrogen bomb ever tested was a 58-megaton Russian bomb in 1961. A 40-million-megaton blast defies comprehension.

The explosion expelled the atmosphere into bright plumes of gas and debris, which at their peak towered 2,200 km above the clouds. After a few minutes, the plume plunged back into the atmosphere. The collapsing plume and the underlying atmosphere became hot and started releasing great fireballs of infrared rays. The point of impact turned into a dark scar, dubbed 'black eye' by astronomers, thousands of kilometres across.

This scary scenario was repeated 20 times, every seven hours, on average. The 'string of pearls' had been reduced to a chain of smouldering scars encircling the planet. Some fragments created bigger bangs than others. The largest of the fragments, a solid body 3 km across, left a 'black eye' of debris as large as Earth. Some of the black eyes took months to fade from Jupiter's face.

The diameter of Jupiter is 11 times greater than that of Earth; a piece of S-L 9 was only about a kilometre wide. Had it struck Earth it would have gouged a crater 60 km across. Comparing S-L 9 with Jupiter is like comparing a fly with a dinosaur. If such a puny crash could cause so much damage, what would happen if a large comet hit Earth?

Did a comet strike Earth in 1908?

The Tunguska fireball was not a meteorite but a small gaseous comet which had left no trace of itself after the impact. That's what two astronomers concluded independently in 1934. F.J.W. Whipple, head of the Kew Observatory in London, said that in view of the fact that the recorded observations of the phenomenon such as bright nights and airwaves were confined to the north of Europe, 'the meteor was essentially a small comet and that the tail of the comet was caught by the atmosphere'. However, he was quick to admit, 'I do not feel much confidence in this hypothesis'.

I.S. Astapowitsch of Leningrad State University expanded the idea suggested by Academician Vladimir Vernadsky, Kulik's mentor, that the Tunguska meteorite could have been a rather dense cloud of cosmic dust, possibly associated with a comet.

Astapowitsch asserted that the bright atmospheric phenomena following the blast could have been caused by the dust tail of the nucleus of a small comet rushing towards Earth and exploding with tremendous energy. 'The explosion gave rise to seismic and airwaves,' he said, 'while the high-temperature explosive waves caused the uniform scorching over the first few kilometres.' He also estimated the blast's energy to be between one and two megatons, about 100 times higher than Whipple's calculation (current estimates vary from 10 to 20 megatons).

In 1942 William H. Christie of the Mount Wilson Observatory in California analysed all available data on the Tunguska explosion. He concluded that the data agreed with the action of a comet:

> The night glows, I think, remove all the doubts as to the nature of the object which struck Earth. It was, apparently, a small comet ... and the tail, or part of it, which was captured by Earth formed the glows. The Greenwich time of the collision was just past midnight; hence no glows were observed much west of Greenwich because this hemisphere would be on the side away from the comet ... The Siberian blast appeared to be the only recorded instance of the collision of Earth with a comet. How long we will have to wait for another such visitor, we do not know, but let us hope it again chooses a sparsely inhabited region of this globe of ours for its final resting place?

At that time scientists knew very little about the structure of nuclei of comets. They believed them to consist of one or a few large stony rocks or even a 'sandbag' of small particles. The size of nuclei was also over-estimated to be several hundred kilometres. Astapowitsch and Whipple's comet hypothesis also turned out to be as questionable as Kulik's meteorite hypothesis, and scientists soon dumped it. It was revived in the late 1950s when the American astronomer Fred Whipple proposed his 'dirty snowball' model for comets.

A comet is a unique cosmic phenomenon: it suddenly appears in the sky, it blazes for a few days, it wows earthlings, it disappears. A comet's awe-inspiring spectacle has always intrigued people. Its irregular appearance in the sky, its varying size, form and brightness, its exotic tail, its abrupt disappearance—these were the mysteries which ancient people were unable to solve. To them, comets were omens of disaster. According to the Greek poet Homer, who lived in the 8th century BC, a comet (from the Greek word *kome*, meaning 'hair') was 'a red star that from his flaming hair shakes down disease, pestilence and war'. In Shakespeare's *Julius Caesar*, after seeing a comet, Caesar's wife Calpurnia warns him: 'When beggars die, there are no comets seen; the heavens themselves blaze forth the death of princes.' Comets are no longer considered harbingers of doom, but they still intrigue astronomers.

Even now, astronomers know very little about comets, and they can't predict when the next one will come. Comets are chunks of matter left over from the birth of the solar system 4,600 million years ago. There are somewhere between two trillion and five trillion comets that circle the solar system in a halo-like cloud—the Oort cloud—between 20,000 and 100,000 astronomical units from Sun (one astronomical unit is the distance between Earth and Sun, about 150 million km). In the Oort cloud, comets are not packed like sardines—the neighbouring comets are typically tens of millions of kilometres apart. The Oort cloud, sometimes called the Siberia of comets because of its freezing temperatures as low as $-270°C$, is named after the Dutch astronomer Jan H. Oort, who in 1950 suggested the existence of a spherical reservoir of comets swirling around the solar system.

At about the same time, Gerard P. Kuiper of the University of Chicago suggested that there is another reservoir of comets, now known as the Kuiper belt. The Kuiper belt extends between 35 and a few hundred astronomical units from the Sun, beyond the orbit of Neptune. It is like a compact disc (CD)—the hole covers the solar system. This CD contains some 200 million 'comet songs' waiting in dusty cold storage for a chance to whoosh across our skies and strike terror into the heart of earthlings.

About a dozen stars pass within about 200,000 astronomical units of the Sun every one million years. Occasionally these passing stars push a comet out of its orbit, sending it towards the inner solar system. It may pass the solar system once and never return, or it may settle into an orbit to visit us regularly, like Halley's comet. Comets that take 20 or more years to orbit Sun, including Halley's, come from the Oort cloud; comets with shorter orbital periods start their journey towards Sun from the Kuiper belt. Orbits of some comets have periods exceeding one million years.

Balls of ice and dust

Comets are fossils—frozen relics from the time of the infant sun. By studying them, astronomers can find out how the Sun and the planets were born. In the early 1950s, Fred Whipple and other astronomers provided an insight into the structure of comets. Whipple said that a typical comet has three parts: a frozen central part called a nucleus, a fuzzy cloud surrounding the nucleus called a coma (or head), and a tail consisting of gas and dust. The nucleus, usually only a few kilometres across, is a 'dirty snowball' made of grains of frozen mass consisting of water, methane, ethane, carbon dioxide, ammonia and many other gases.

In 1986 the European Space Agency's Giotto spacecraft proved that Whipple's model was reasonably accurate when it took close-up photographs (from a distance of 480 km) of the nucleus of Halley's comet. In essence, a comet nucleus resembles a fluffy snowball coated with a crust of black material and spouting jets of vaporised ice.

As the comet approaches the Sun, the gases evaporate under the Sun's heat and form its coma, which scatters sunlight. The solar wind, a spray of charged particles from the Sun, blows this material into a tail, reaching typical lengths of 10 million km, and in rare cases, several times that distance. The tail is always away from the Sun. Comets become visible only when they are close to Sun—between two and three times as far away from the Sun as Earth. That's why astronomers can't be sure when the next new comet will come.

Whipple, often referred to affectionately as 'Dr Comet', not only coined the evocative phrase 'dirty snowball', he also came up with an equally evocative idea: a comet is like a jet engine. Like the heated gases erupting from a jet engine, the evaporating gases from the nucleus exert a force on the nucleus. This force gives the comet its independent thrust. 'When I first realised the jet action of comets,' 79-year-old Whipple told *Time* magazine in 1985, 'Boy! That was a thrill.'

The most famous of all comets is Halley's comet. It's named after the 18th-century British astronomer Edmond Halley, who first calculated its period and successfully predicted its return in 1758. It travels in a giant orbit that takes about 76 years. It was last seen in 1986 and is expected to return in 2061. It's a long wait for Halley's, but we never know—we may be lucky enough to see one night a brilliant comet arcing through the sky trailing a magnificent tail. It will outshine everything in the sky. When Halley's comet was here in 1910, Earth passed through its tail without any effect. It is a different matter if a comet head hits Earth.

Comets usually pass Earth with speeds greater than 160,000 km/h. If a comet a couple of kilometres across hits Earth with such a speed, it will gouge out a hole as big as a large city, spewing out so much dust in the atmosphere that the Sun will be blocked out for months. If it hits an ocean, a tidal wave up to one kilometre high and travelling at many hundreds of km/h will submerge most low-lying regions. But the worst threat is fire—fire caused by debris thrown into the atmosphere when the comet head explodes before hitting Earth. Burning forests and cities will throw soot into an already clogged atmosphere. Then there will be acid rain—a rain of toxic gases and metals. Most plants will die, followed by marine creatures that live near the surface, nearly ending life on the planet.

The comet did it

Back to Tunguska. The scary scenarios of a comet striking Earth are hypothetical, but the devastated Siberian taiga is a reality. The Soviet Academy of Sciences was determined to find an answer, and in 1954 sent Kirill P. Florenskiy, a geochemist, to survey the Tunguska site. Using modern cartographic equipment, Florenskiy made an aerial survey of the region and drew up a new, more accurate map. In 1957 the Russian mineralogist A.A. Yavnel microscopically analysed soil samples brought back by Kulik in 1929 and 1930. In some samples, he discovered magnetite globules 30 to 60 micrometres in size. The samples also showed a small percentage of nickel and traces of cobalt. The results suggested that an iron meteorite had fallen in Tunguska.

These results encouraged the Academy to organise a new expedition. In the summer of 1958, Florenskiy headed the fifth Tunguska expedition, the first after the Second World War. The main aim was not to look for visible meteorite fragments but to collect micrometeorites in the soil. The expedition failed to confirm the existence of magnetite and nickel globules detected by Yavnel. It was discovered later that Kulik's samples had been contaminated with samples from the iron meteorite that fell in the Sikhote-Alin mountains in Siberia on 12 February 1947.

The soil samples from this iron meteorite—which destroyed an area of more than a square kilometre and created many craters, the largest being 26.5 m in diameter and 6 m deep—had been stored with Kulik's samples. In his report to the Academy, Florenskiy rejected Kulik's hypothesis that the explosion took place on the ground and suggested that it had taken place in the air some height above.

Florenskiy headed another expedition to Tunguska in 1961-62, which analysed in detail the pattern of the devastation of the forest and established the trajectory of the fireball. However, by 1960 Florenskiy was convinced that Tunguska had been flattened by a comet. 'Many facts favor the view that the colliding body was a comet: the unusually loose structure, which led to breakup in the atmosphere; the dust tail, pointing away from Sun, which caused unusual sunsets over nearly all over Europe; the nature of the orbit; and lack of big fragments,' he wrote in *Sky & Telescope* magazine. He pinpointed the location of the explosion at an altitude of 5 km, southeast of the centre of destruction on the ground, and suggested that the collision was neither head-on nor the comet directly followed Earth, for it struck Earth almost squarely on the side.

He explained the explosion in Earth's atmosphere by saying that the sudden stopping of a body moving at 18,000 km/h releases enough heat to vaporise it instantaneously. If the body is loosely compacted and contains volatile matter such as gases or ice, deceleration in the air may cause explosive vaporisation without the body striking the ground, he suggested.

Florenskiy established that after the explosion the Tunguska forest burned for at least five days. A weak southeasterly wind at 7.2 to 18 km/h was blowing on that day, but the fire was preceded by a dry spell and therefore spread rapidly through the treetops. The fire died out because of the unfavourable weather conditions.

'We found that branches which had been not thicker than a fountain pen before the catastrophe still retained marks of their injuries,' he said. 'The damage is noticeable on the upper portions of the branches, thus making it impossible to associate these injuries with ordinary fires. The severity of these injuries diminished significantly with increasing distance from the epicentre.'

He estimated that the energy required for the observed injuries would be 5 to 15 calories per square centimetre. This value could not be significantly higher since this would lead to marked charring of the bark, and no such phenomenon was observed. 'Approximately the same energy is required to ignite dry forest debris, and this could lead to forest fire,' he noted. Studies of nuclear explosions have shown that a 10-megaton explosion would produce charring similar to the Tunguska forest in fir, pine and maple bark.

In the early 1960s, Vassilii Fesenkov of the Soviet Academy of Sciences' meteorite committee expanded the comet theory. He gave various reasons in support of his belief. These can roughly be grouped into four categories:

First, craters found by Kulik are now not considered to be places of the fall of fragments of the meteorite. Despite extensive searches, no primary meteorite fragments have been found. It's easier to visualise a 'dirty iceberg' exploding to nothing than a rocky meteorite. Many magnetite and silicate globules (5 to 450 micrometres in diameter) found in the area were 'clearly of secondary origin'. They were most likely formed in the atmosphere due to rapid condensation of molten 'raindrops' as they drifted to the ground. The height of the comet's explosion—5 to 6 km above Earth's surface—has been well established by the measurement of shock waves received at observatories in Irkutsk (Siberia) and Potsdam (Germany) and the six microbarograph stations in England.

Second, according to all evidence, this meteorite moved around the Sun in a retrograde direction—it was moving from south to north at a time when Earth was generally moving from north to south—which is impossible for typical meteorites. Meteorites rarely hit Earth in the morning, because the morning side faces forward in the planet's orbit. Usually, the meteorite overtakes Earth from behind, on the evening side. Comets, on the other hand, have a wide range of orbits and speeds and could collide with Earth on the morning side, hitting head-on with a speed of about 145,000 km/h. The Sun's glare prevented any sighting of the comet before it hit Earth because its direction and the angle of strike towards Earth were from behind the Sun.

Third, the most probable explanation for the brilliant night skies observed in the British Isles, Europe and Western Russia would be that the meteorite was a little comet with a dust tail pointing away from the Sun. The dissipation of this tail in the atmosphere significantly increased the night sky brightness. The dust particles causing the night glows were at the height of several hundred kilometres above Earth's surface and did not behave as ordinary meteor showers.

Fourth, a marked decrease in the air's transparency, recorded two weeks after the explosion, was caused by the loss of several million tonnes of dust from the comet during its flight through the atmosphere.

Fesenkov told *The New York Times* on 20 November 1960 that 'the recent study suggests that explosions took place over at least three points in the area'. He said that in one of the Soviet studies, an experimental model was used to simulate the forest and miniature explosions were set off in the air. These experiments suggested that the head of the comet, a 'small one,' consisted of dust and frozen gas in one or more extremely dense clouds several kilometres in diameter.

'The total weight is thought to have been more than 1,000,000 tons,' The *Times* reported. 'When it hit the atmosphere, the resulting explosion is thought to have been comparable in force to that of an equal amount of TNT.' It's worth noting that Fesenkov's estimate of the blast's energy—one megaton—is similar to the estimate made by Astapowitsch in 1933.

Further support for the comet theory came in 1975 from Ari Ben-Menahem, an Israeli scientist, who re-analysed the old seismographs of the Tunguska explosion and compared them with a series of air explosions from nuclear tests at the USSR test site at Novaya Zemlya. He concluded that the explosion took place 8.5 km above the ground and had the energy of about 12.5 megatons.

A year later, David Hughes, a British astronomer, estimated the comet's nucleus diameter at about 40 m, much smaller than the diameter estimated for visual comets. The small diameter explained why the comet wasn't seen as it approached on its collision course to Earth. 'A cometary nucleus of this size will hit Earth about every 2,000 years, the rarity of the event giving ample justification for visiting Tunguska again,' he said.

In a further study, Hughes and his colleague John Brown noted that, although the temperature produced by the burn-up of the comet in the atmosphere would have been no more than a few million degrees, too low for nuclear reactions, this temperature was high enough to produce nuclear-like effects such as the production of X-rays, gamma rays and highly accelerated charged particles. Even if the Tunguska body did cause nuclear effects, it does not mean it was not a comet. 'The Tunguska explosion was by an impacting small comet and nothing more exotic needs to be invoked,' Hughes and Brown concluded.

After a more recent analysis of the unusual sky glows, Vitalii Bronshten of the Committee of Meteorites of the Russian Academy of Sciences concluded that 'the cause of the glow was secondary scattering of sunlight by the dust constituting the envelope of the Tunguska comet'. Bronshten, one of the main contemporary supporters of the comet theory, first calculated the volume density of the dust ejected by the nucleus of a comet like Halley's at a distance of several thousand kilometres, and then its transfer to the west in Earth's gravitational field, considering deceleration caused by the atmosphere. He showed that the bigger dust would reach the British Isles in six hours. Smaller particles could cover greater distances, but they could not produce a noticeable scattering of light. This is the reason why the glowing area had a western border: night glows were limited to the British Isles, Europe and Western Russia.

If the Tunguska object was indeed a comet, then it must be a comet known to astronomers. In 1978 the Slovak astronomer Lubar Krésak suggested that a piece of comet Encke had exploded at Tunguska. He based this idea on the fact that the fireball exploded at the peak of one of the most intense annual daytime meteor showers in late June, which has long been thought to derive from this comet. Encke's comet is named after the mathematician who investigated its orbit. The German astronomer Johann Encke was born five years after the comet was discovered in 1786, but he showed in 1822 that it has a period of three years and four months, the shortest known.

Was it a 'fair dinkum' comet?

In 1975 Fred Whipple questioned the possibility of a comet striking Tunguska. He estimated that if we take the mass of the comet as one million tonnes, as suggested by Fesenkov, the chance of such a comet striking Earth every 100 years would be about one in 20,000. 'It appears unlikely, therefore, that the Tunguska explosion was produced by a bona fide active comet a hundred or so meters in dimension … more likely, however, the Tunguska object was an inactive, low-density, friable body … There is no reason to suspect it was interstellar.'

Most contemporary scientists also reject the idea of a comet. 'Comets are fluffy in comparison with asteroids and burn up quickly in the atmosphere,' Richard Stone writes in *Discover* magazine. 'For one to have produced an explosion as big as the one over Tunguska, it would have started out as a million-tonne object. The vast swath of gas and dust left by such an object on its way down might well have shut out Sun or altered the climate'. Zdenek Sekanina, an expert on comets at NASA's Jet Propulsion Laboratory, agrees: 'The effect on life on Earth would have been horrendous. It would have been a global catastrophe, comparable to nuclear winter. The effects on humanity would have been so overwhelming that we could not discuss the topic because we would not be here'. You and I are still discussing the topic; therefore, the Tunguska fireball was not a comet. QED.

Although the probability of such low-density objects colliding with Earth is quite small, there is still hope for the comet theory—and it comes from Down Under. Assuming a speed of 108,000 km/h, as the Australian scientists Duncan Steel and Richard Ferguson present their case, seven hours before the impact the Tunguska object would have been about 750,000 km from Earth. Active comets produce tails that stretch millions of kilometres away from the Sun, so it is possible that there could have been an encounter between Earth and the tail of the comet. This encounter could produce an aurora in the hours before the impact. But did anyone observe an aurora seven hours before the Tunguska blast?

Steel, a well-known authority on the threat posed by asteroids, was at a conference on asteroids, comets and meteors in Sweden in 1989, when the Russian scientists Nikolai Vasiliyv and G. Andreev circulated a short report on Soviet research on Tunguska. He was intrigued by the following paragraph in the report:

> A unique item in this respect could be the search for the original diary entries by Mouson who observed auroras from near the Erebus volcano at Antarctica during the summer of 1908. There is information in Shackleton's accounts that on June 30, Mouson registered an aurora which he visually considered to be anomalous. Unfortunately, Shackleton's accounts do not contain further details.

Steel soon figured out that 'Mouson' was 'Mawson,' after transliteration from the Latin script to Cyrillic and then back again, and 'the summer of 1908' was in fact 'the Antarctic winter of 1908.' Coincidentally, at that time Steel was at the University of Adelaide, where Mawson's notebooks from the Antarctic Expedition of 1907-09, led by Sir Ernest Shackleton, are archived in the Mawson Institute for Antarctic Research.

Mawson (later Sir Douglas Mawson) was a young geologist on the expedition, and kept a diary of his observations. Steel and Ferguson made extensive searches of Mawson's diaries and all other expedition papers but failed to find any record of aurora australis at the time of the Tunguska blast. However, they found a record of an exceptional aurora seven hours before the blast.

Did the Tunguska fireball cause this aurora? If yes, then the fireball was, as Mawson would have said, a fair dinkum (genuine) comet.

Comet theory revived

In 2010 Vladimir Alexeev of Russia's Troitsk Innovation and Nuclear Research Institute visited Tunguska to examine the famous Suslov crater with GPR, an instrument capable of exploring the ground to the depth of 100m. After the visit, he concluded that the Suslov crater was formed by a gigantic piece of ice that broke away from the nucleus of a comet. The fragments of this piece of ice wreaked havoc on Tunguska trees. The tail of the comet scattered dust in the atmosphere, causing strangely glowing skies. Alexeev is convinced that the old comet theory is indeed the right theory.

According to Duncan Steel, the chunk was from a massive comet that has been disintegrating for tens of thousands of years, and this disintegrating comet would have been a threat even to prehistoric humans. Its impact risk peaked around 5,000 to 5,500 years ago when the work first began on Stonehenge, the most important of all prehistoric monuments. Steel's calculations suggest that it was initially built as an ancient early warning system for impending impacts by Tunguska-like objects. On the day of greatest danger, the comet debris would have been visible just above the horizon in line with the 'Heel Stone' set outside the main circle, he says. A fascinating idea, indeed.

Chapter 4

Asteroids Behaving Badly

Once called boring 'vermin of the skies,' asteroids are now star attractions for astronomers around the world. This attention is worthy of their name, which is Greek for 'starlike', but it has more to do with their sheer number and their destructive power than their cosmic beauty. These pockmarked giant peanut-like rocks are in fact leftovers from the formation of the planets.

The largest three asteroids are Ceres, Pallas and Vesta, with average diameters 930, 520 and 500 km respectively. About 200 asteroids are larger than 100 km across, 800 larger than 30 km. About a million are one kilometre or more in diameter, and billions are of boulder or pebble size. Most of the asteroids orbit within a vast, doughnut-shaped ring between Mars and Jupiter, known as the main belt.

Occasionally, a collision may kick an asteroid out of the belt, sending it onto a dangerous path that crosses Earth's orbit. These stray asteroids take up an orbit that loops past Earth and are called 'Earth-crossers'. This knowledge frightens astronomers. What if one of them comes too close to Earth? What cataclysm would such a rogue rock cause if it slams into Earth? The number of asteroids is very large, but the space they occupy is enormous. Most asteroids stay millions of kilometres apart. It's not like Star Wars or Star Trek spaceships weaving their way through flying rocks. But real collisions are possible with a spacecraft or Spaceship Earth.

In 2002 NASA's Jet Propulsion Laboratory (JPL) established its Sentry program This highly automated collision monitoring system continually scans the most current asteroid catalogue for possibilities of impact with Earth over the next 100 years. Whenever a potential impact is detected, it is analysed, and the results are immediately published on the website of JPL's Centre for Near Earth Objects Studies: cneos.jpl.nasa.gov/sentry/

So far NASA has discovered about 20,000 Earth-crossers, or near-Earth asteroids, of which 863 are more than a kilometre wide and 1,379 are labelled as 'potentially hazardous' to the planet. Don't panic—none of them is on a collision course yet. On 15 February 2013, a 45-metre-wide asteroid, called 2012 DA14, came within 27,000 km of Earth. For you and me, 27,000 km is, well, a long way away. For astronomers who coin words like 'Earth-crosser' it is a hair's-breadth; for us, a hair-raiser indeed. If this 130,000-tonne rock were to hit, it could destroy a city twice the size of London.

The NASA's Sentry program also keeps track of the asteroid 1950 DA. As the name suggests, it was discovered in 1950. It was observed for 17 days but then it faded from view for half a century. It was rediscovered on New Year's Eve 2000. There is a one in 300 long shot that the 1.1-kilometre-wide 1950 DA would hit Earth in March 2880. Don't forget to pass the JPL website's good news on to your next 35 generations: 'There is no reason for concern over 1950 DA. The most likely result will be that St Patrick's Day parades in 2880 will be a little more festive than usual as 1950 DA recedes into the distance, having passed Earth by.'

Another asteroid, dubbed 2020 DR2, measuring about 590m in diameter, has a small chance of hitting Earth either on 10 September 2074 or 9 September 2081. If this 280-million-tonne and 590-m-wide monster strike the planet, the force of impact will cause mass causalities worse than any natural disaster in human history.

Did an asteroid a strike indeed take place on 30 June 1908?

Pointing an accusing finger

After meticulously reviewing five decades of research on the Tunguska fireball, in 1983 Zdenek Sekanina, a Czech-American astronomer, put together a new analysis of the orbit, atmospheric entry and explosion of the interstellar body. He ruled out a comet as a suspect and pointed an accusing finger at an asteroid about 90 to 190m across. The asteroid came in from a direction close to 110 degrees east of north at an angle about 5 degrees above the horizon and exploded at about 8 km above the ground.

Its speed when it entered the atmosphere was about 108,000 km/h.

In his 32-page seminal paper in *The Astronomical Journal*, Sekanina remarked that evidence on the forest devastation in the area of fall of the Tunguska fireball left no doubt that it exploded in mid-air and then completely disintegrated in the atmosphere. The evidence included the absence of impact craters and/or sizeable debris in the area of the fall, near-perfect radial symmetry of the region of flattened trees to within at least 15 km from its centre, and the presence of standing bare trunks (Kulik's 'telegraph poles') at the very centre of the radial of devastation.

Since the evidence pointed to only one enormous outburst, Sekanina concluded that the fireball did not fracture in flight. If such break-ups had occurred, the resulting sequence of explosions would have diminished the enormous power of the final explosion.

The explosion released almost instantaneously energy above 12 megatons. The eyewitnesses saw this energy, which was enough to wipe out even modern London or Tokyo, as a fireball 40 times brighter than the noonday sun. 'This conclusion is supported by accounts of a dazzling blaze of light in the sky, described in eyewitness reports from places near the centre of the explosion,' Sekanina said.

The fireball dissipated a million tonnes of small particles in less than one-tenth of a second in Earth's atmosphere. Winds assisted by the expanding shock front dispersed the dust in the stratosphere, which caused the bright skies reported in the aftermath of the fall.

Sekanina rejected the notion that the Tunguska fireball was a comet. Because of the high speed at which the object was travelling when it entered the atmosphere, it resisted a very high air pressure before exploding. It is inconceivable that a comet, known for its extreme fragility, could have survived such high pressure. The fireball, therefore, must have been a denser, stony object to survive its journey to the Siberian sky. It was probably a small member of the Apollo asteroids, believed to be nuclei of comets that have lost their volatile components.

Critics of the asteroid theory said that Sekanina's analysis was mainly based on eyewitness accounts, most of which were recorded at least two decades after the event, and seismic records. 'You can't make a sophisticated model from poor data,' the American meteor specialist Richard McCrosky told *Sky & Telescope*. 'It seems everything he assumes must be true for his conclusion to be right.'

In the same year, the American atmospheric scientist Richard Turco suggested that the bright nights were caused by noctilucent clouds, silvery clouds at high altitude that shine at night. On 30 June 1908, the wind was blowing in the right direction for the dust associated with the fireball to reach western Europe. The dust would then have settled in the atmosphere at an altitude of about 80 km and remained there for several days. These dust particles and the water vapour that was also deposited by the fireball contributed to enhancing cloudiness. A stony asteroid would have been too dry to provide the water needed for such clouds. Turco's analysis favoured an icy comet nucleus.

More recently, Vitalii Bronshten rejected the asteroid theory on the ground that an analogy with nuclear explosions suggested that, even upon a powerful explosion, a stony asteroid should break up into fragments of various sizes. These fragments could not completely vaporise in the intense heat of the explosion or as a result of their fall to Earth. Some of these fragments should survive, but even after many careful searches, no such fragments have ever been found. Bronshten also suggested that the noctilucent clouds, which he agreed were observed, could never gain such brightness to result in light nights.

An asteroid masquerading as a planet

My Very Educated Mother Just Served Us Nine Pizzas—the mnemonic some of us learned at school to list the order of the nine planets (outwards from Sun) is now without our favourite Pizza for Pluto. In 2006, demoting Pluto, the International Astronomical Association recommended that your mother should serve nachos instead: My Very Educated Mother Just Served Us Nachos. At its general assembly in Prague, the IAU defined a 'planet' as a celestial body that:

a. is in orbit around the Sun;
b. is massive enough to have its self-gravity to pull itself into a round (or nearly round) shape; and
c. there are no other bodies in its path that it must sweep up as it goes around the Sun.

Pluto fails to meet criterion (c). For in about twenty years, Pluto's 248-year orbit, will bring it closer to the Sun than Neptune.

Pluto isn't rocky like Mercury, Venus, Earth and Mars, nor is it a gas giant like Jupiter, Saturn, Uranus and Neptune. It is a relatively tiny ball of ice; with a diameter of 2,274 km, it is smaller than the Moon. Pluto's orbit is also peculiar. The other outer planets orbit Sun in roughly circular orbits, but Pluto's orbit is elliptical, which at times brings it closer to Sun than Neptune. The oddball Pluto is not a planet—that's what some astronomers have been arguing for years. It's the largest object in the Kuiper belt, the outpost of icy asteroids beyond the orbit of Neptune.

The existence of such a belt was first suggested in the early 1950s but the first Kuiper belt object, called 1992 QB1, was found in 1992. This small icy body, similar in size to an asteroid, suggested that there might be more than just Pluto in the distant reaches of the solar system. Since then, hundreds of objects like QB1 have been found in the Kuiper belt. Their diameters range from 50 to almost 1,200 km. Though these objects are smaller than Pluto, they are thought to be similar to Pluto in composition.

The Kuiper belt contains about 100,000 objects larger than 100 km across. This swarm of Pluto-like objects gives weight to some astronomers' argument that Pluto is not a planet but a huge ball of ice. Even those astronomers who defend Pluto's status as a planet agree that were it discovered today it probably wouldn't be called a planet. This debate about Pluto's status as a planet has a historical precedent. More than two centuries ago when Ceres was discovered, it was also proclaimed a planet.

Piazzi's 'planet'

The discoverer was defiant: 'I have the full right to name it in the most convenient way to me, like something I own. I will always use the name Ceres Ferdinandea, nor by giving it another name will I suffer to be reproached for ingratitude towards Sicily and its king.'

Shortly after nightfall on 1 January 1801, Giuseppe Piazzi, a monk and director of the brand-new Palermo Observatory built atop a 12th-century tower in the royal palace of Sicily, pointed his shining brass telescope at stars in the constellation Taurus and observed an unfamiliar, faint, starlike object. Later observations and calculations showed it to be the 'missing planet' between Mars and Jupiter. He named it Ceres Ferdinandea—Ceres for the patron goddess of Sicily, and Ferdinandea for his royal patron, King Ferdinand of Naples and Sicily.

The discovery caused a sensation in Europe. The burning question was: what should the new planet be named? Napoleon even discussed it with the celebrated French mathematician and astronomer Pierre Laplace. Some French astronomers suggested Piazzi, while some Germans favoured Juno or Hera. But Piazzi was adamant that he had the right to name it.

The story of Piazzi's discovery starts in 1772, when Johann Titius, a professor at Wittenberg in Germany, discovered a remarkable numerical relationship between the distances of the planets from Sun. He pointed out that the numbers in the series 0, 3, 6, 12, 24, 48, 96, when added to 4 and divided by 10, produced the series 0.4, 0.7, 1, 1.6, 2.8, 5.2, 10. If Earth's distance from the Sun is set at one astronomical unit (about 150 million km), then these numbers give the distances of the six planets known at the time, except for position 2.8. Titius suggested that this gap belonged to still undiscovered satellites of Mars.

That same year, the German astronomer Johann Bode picked up Titius's rule and quoted it without any acknowledgement in his astronomy textbook. However, he suggested a new planet for the gap at 2.8. The rule is now known as Bode's law. Although Bode also carried out other astronomical investigations, he is remembered today for popularising a relationship that he did not originate.

When the celebrated German-British astronomer William Herschel discovered the planet Uranus in 1781, it also fitted Bode's law (continuing the Titius series by doubling 96 for Saturn, that is, 192; when added to 4 and divided by 10, this gives 19.6, which is close enough to 19.2, the actual distance of Uranus from Sun in astronomical units). Astronomers now strongly felt that another planet was to be found between Mars and Jupiter.

The Hungarian astronomer Franz von Zach so strongly believed in the 'missing planet' that he tried to calculate its orbit by using Kepler's laws, but one element that might reveal its location—the longitude—eluded him. In 1785 he wrote to Bode: 'I am having much the same success as the alchemists in their search for gold; they had everything except the vital factor. Thus I too believe that I am in possession of all the elements of the orbit of this still unknown planet, except one; this alone now keeps me amused, and although one may not find gold in one's wanderings, one does occasionally come across a chemical process.'

In 1787 von Zach took a solo search for the planet, but without success. 'It cannot be a matter for one or two astronomers to scrutinise the entire Zodiac,' he wrote in *Monatliche Correspondenz* (Monthly Correspondence), the world's first astronomical journal, which he founded. The hunt for the missing planet began in earnest when, in 1800, von Zach organised a group of 24 astronomers who called themselves the 'celestial police'. They divided the entire Zodiac into 24 zones. The zones were then allocated to the members by lot. Each member was to be responsible for drawing up a star chart for his zone.

'Through such a strictly organised policing of the heavens, divided into twenty-four sections, we hoped eventually to track down this planet, which had so long escaped our scrutiny—supposing, that is, that it existed and could be seen,' von Zach wrote in *Correspondenz*. Before 'such a strictly organised policing' of the heavens could get under way, surprising news arrived from Palermo on the island of Sicily.

Giuseppe Piazzi, a Theatine monk who entered the order in 1764 at the age of 18, received his early training in philosophy but later in life took up mathematics and astronomy. In 1780 he was called to the chair of higher mathematics at the Academy of Palermo. There he soon obtained a grant for an observatory, and he went to England in 1788 to buy instruments for it. He commissioned Jesse Ramsden, the greatest of the instrument-makers, to make him a 1.5-m vertical circle of unique design for measuring the altitudes and azimuths of stars by micrometer microscopes. While in England he became acquainted with Herschel and had the unique 'privilege' of falling off the high wooden ladder at Herschel's large reflecting telescope and breaking his arm.

Once the Ramsden circle, the masterpiece of 18th-century technology, was installed at Palermo in 1791, Piazzi started his painstaking work of cataloguing stars. In 1803 he published his first catalogue containing 6,784 stars, and in 1814 a second one containing 7,646 stars. However, Piazzi's major accomplishment did not involve the stars at all. It was the discovery of the 'missing planet'.

New century's gift

On 1 January 1801, the first evening of the new century, Piazzi observed an unfamiliar point of light in the sky. He thought that the object might be a new star. Over the next three evenings, he observed it again and noticed that it had shifted its position at the same rate as on the preceding days. Piazzi was now sure that it was not a fixed star. Thinking it might be a comet, he continued to follow it until 11 February when illness cut short his work. However, on 24 January he announced his discovery to Bode, the French astronomer Joseph Lalande, and his friend Barnaba Oriani, director of the Brera Observatory in Milan.

He confided only to Oriani that it might be a new planet: 'I have announced this star as a comet, but since it is not accompanied by any nebulosity and, further, since its movement is so slow and rather uniform, it has occurred to me several times that it might be something better than a comet. But I have been careful not to advance this supposition to the public.'

Oriani replied: 'I congratulate you on your splendid discovery of this new star. I do not think that others have noticed it, and because of its smallness, it is unlikely that many astronomers will see it'. Bode thought that Piazzi's discovery marvellously fulfilled his prediction of a planet between Mars and Jupiter. Von Zach was elated and reported the news in *Monatliche Correspondenz*, under the heading 'On a long supposed, now probably discovered, new major planet of our solar system between Mars and Jupiter'.

But Bode and von Zach could not verify the discovery. Such was the state of the postal service in those days that Bode did not receive Piazzi's letter until 20 March. By now the new planet had ceased its retrograde motion and had begun to advance and had moved near enough to Sun that it could not be seen. Everyone eagerly awaited its emergence from the other side of Sun in July. Herschel was the first to search for it in July, and he and many other astronomers continued their search for months. The planet was missing again.

It required the genius of the German mathematician Carl Friedrich Gauss to recover Piazzi's lost planet. Gauss, then 24 and at the beginning of a brilliant career that placed him in the company of Archimedes and Newton, calculated the orbit of the planet from Piazzi's few observations. Gauss's calculation of the planet's position was so accurate that on 31 December 1801, within a few hours of one another, von Zach and the German amateur astronomer Heinrich Olbers independently recovered the lost planet.

Piazzi named it Ceres Ferdinandea, but it was soon shortened to Ceres. King Ferdinand wanted to strike a gold medal with Piazzi's effigy, but the astronomer requested the privilege of using the money to buy a much-needed equatorial telescope for his observatory.

The discovery of Ceres posed a problem for astronomers. Herschel's observations showed that it was a most unusual planet, being too small to show a planetary disc. The problem was compounded when, on 28 March 1802, Olbers discovered 'another Ceres,' a body that was also orbiting in the 'gap' between Mars and Jupiter. Olbers, a medical doctor by profession, named it Pallas. The discovery was most perplexing.

Olbers wrote to Herschel: 'Could it be that Ceres and Pallas are just a pair of fragments, or portions of a once greater planet which at one time occupied its proper place between Mars and Jupiter and was in size more analogous to the other planets, and perhaps millions of years ago, either through the impact of a comet, or from an internal explosion, burst into pieces.'

Within a month of the discovery of Pallas, Gauss calculated its mean distance from the Sun, almost the same as Ceres. He also noted that Ceres and Pallas had many characteristics that made them quite unlike planets. Bode was not yet convinced that his law (which he had unceremoniously pinched from Titius) was not true. He wrote to Herschel: 'I hold myself convinced that Ceres is the eighth primary planet of our solar system and that Pallas is a special exceptional planet—or comet—in her neighbourhood, circulating around Sun. So there would be two planets between Mars and Jupiter, wherever since 1772, I have expected only one; and the well-known progressive order of the distances of the planets from Sun, is by this fully proved.'

Herschel was not influenced by Bode's plea for his law. He was now convinced that Ceres and Pallas represented a new and different class of celestial bodies. He also believed that since Ceres and Pallas did not occupy the space between Mars and Jupiter with 'significant dignity,' these new bodies were not worthy of the name 'planet'. He proposed that they should be given the name 'asteroid' (from the Greek word *asteroeides*, meaning 'starlike'), since they are intermingled with, and similar to, the small fixed stars. He went on to advocate three forms of celestial bodies—planets, asteroids and comets.

Most astronomers now accepted that Ceres and Pallas were not planets, but Piazzi was not happy with Herschel's celestial hierarchy of planets, asteroids and comets, and retorted: 'Soon we shall be seeing counts, dukes and marquesses in the sky.' He suggested the name 'planetoids', pointing out that 'asteroid' would be more appropriate for 'little stars'. The term 'asteroid' has persisted, but they are sometimes also referred to as 'planetoids' or 'minor planets'.

Von Zach's 'celestial police' did not give up their quest and continued to scan the heavens through their telescopes. Their efforts were rewarded when German astronomer Karl Harding discovered the third asteroid,

Juno, on 1 September 1804. Olbers added the fourth, Vesta, on 29 March 1807. Both were too tiny to be qualified as a planet. The talented Olbers also suggested an idea which has passed the test of time: noting that the brightness of Ceres and Pallas appear to vary from one observation to another, he said that asteroids have an irregular rather than round shape. To Olbers, asteroids appeared like rocks tumbling through space. And he was right. However, he is now best known for the Olbers paradox, the answer to the deceptively simple question: why is the sky dark at night?

The thousandth asteroid was discovered in 1923 and was named Piazzia in honour of Piazzi (who had died nearly a century earlier in 1826 at the age of 80). Since that time, no year has passed without the discovery of new asteroids. Now discoverers do not have to fight for their right to name their discovery. Once the precise orbit of an asteroid is determined, it is given a permanent catalogue identification consisting of a number that denotes its order of entry, which is usually followed by a name proposed by the discoverer, for example, 1 Ceres, 2 Pallas.

Until the discovery is approved by the International Astronomical Union, the asteroid is provisionally known by its year of discovery followed by two letters and numbers, if necessary, indicating the date and sequence of discovery. For example, 1950 DA—the first letter 'D' signifies that it was discovered in the period 16-28/29 February, and the second letter 'A' shows that it was the first discovery during that period.

Target Tunguska

In 1993 NASA scientists Christopher Chyba and Kevin Zahnle and their colleague Paul Thomas from the University of Wisconsin gave the asteroid theory for the Tunguska explosion new weight and rigour.

Their computer simulations to explain the pattern of trees blown down showed that the explosion released about 15 megatons of energy in the atmosphere at an altitude of about 8 km but did not crater Earth's surface. They then examined the entry of three classes of asteroids (stony, iron and carbonaceous) and two classes of comets (short-period and long-period) starting with 15 megatons of kinetic energy. Their simulation showed that cometary nuclei and carbonaceous asteroids explode far too high in the atmosphere to account for the blast, and iron asteroids hardly fragment and so hit the ground at high speed. Only the stony asteroids would create a Tunguska-like explosion.

In their innovative analysis, the researchers included the effects of aerodynamic forces on an asteroid: as the asteroid moves deeper into the atmosphere, atmospheric drag on it increases. When the drag exceeds the asteroid's strength, the asteroid crumbles and begins to flatten like a pancake. The increasing surface of the fragmented asteroid experiences a sharper rise in drag. Increasing drag slows down the asteroid, which in turn spreads it even more. At the same time, atmospheric density rises with decreasing altitude, creating more drag. These increasing forces stop the asteroid abruptly in the atmosphere. The asteroid explodes like a bomb, and within a fraction of a second megatons of kinetic energy is converted into high temperatures and high pressure. The asteroid vaporises. A shock wave races outwards.

Taking account of these effects, the researchers calculated that a stony asteroid about 30 m in diameter and moving at 54,000 km/h would explode at a height of about 8 km, the same height at which the Tunguska body apparently exploded. A smaller one would have exploded much higher, and a larger one would have created an impact crater.

The weak, fast-moving, easily crumbled comets do not penetrate the atmosphere deeply and are unlikely to approach the altitude of the Tunguska explosion. 'Even if the comets had strengths comparable to stony asteroids, they still could not fit the Tunguska observation,' the researchers maintained. For example, if a 15-megaton incident comet is assigned an anomalously low asteroid-like speed of 54,000 km/h, it would completely exhaust itself before it reached an altitude of about 16 km. It would have also caused far less surface destruction.

The measurements of Halley's comet by the Giotto spacecraft in 1986 showed that it had a density of between 0.6 and one gram per cubic centimetre. In their simulation the Chyba trio used a density of 1 but said that values as low as 0.3 might be possible. Such lower-density objects would airburst even higher. 'Tunguska was probably a fairly strong, dense, asteroid-like object, but probably not as strong or dense as iron,' the researchers concluded in their report in Nature. 'Carbonaceous asteroids and especially comets are unlikely candidates for the Tunguska object.' However, their simulation did not completely rule out an unusually fast iron asteroid or a very strong carbonaceous asteroid.

Chyba's team also supported Turco's 1983 observation that bright nights were caused by noctilucent clouds. They said that the air heated by the explosion injected enough water into the stratosphere for noctilucent clouds to be produced.

Further support for the asteroid theory came from Jack Hills and Patrick Goda of the Los Alamos National Laboratory. In their general study of fragmentation of asteroids, they found that a stony asteroid must be greater than 200 m across, much larger than the Tunguska object, to hit the ground. They agreed that the Tunguska evidence ruled out a comet, and that furthermore there is no question but that the object was a stony asteroid and not an iron asteroid. Comets start fires more easily than asteroids, but the Tunguska asteroid generated enough heat to ignite pine forests. 'However, the blast wave from an impacting asteroid goes beyond the radius in which the fire starts,' they said. 'The blast wave tends to blow out the fire, so it is likely that the impact will char the forest (as in Tunguska), but the impact will not produce a sustained fire.'

Henry J. Melosh of the University of Arizona in Tucson, commenting on the Chyba team's research, said that they had 'wrapped up' the most believable explanation for the Tunguska event. 'Substantial progress has thus been made in reducing the Tunguska explosion from the realm of the near-miraculous to a natural, although rare, occurrence,' he said.

Chyba declared in *Astronomy* (December 1993): 'According to this picture, Tunguska goes from an exotic event demanding the invocation of UFOs or black holes to a completely normal, predictable outcome for a stony asteroid entering the atmosphere with a typical velocity. This understanding is important, for it allows us to assess the contemporary hazard posed by small asteroids and comets colliding with Earth.'

Fifteen years after proposing his original asteroid theory, in 1998 Zdenek Sekanina revisited his earlier analysis and concluded that the interpretation of the Tunguska event as a fall of a small asteroid is 'not only plausible, but virtually certain'. However, he said that one issue where he was not prepared to take sides was whether the object was a stony or a carbonaceous asteroid.

In his new analysis, Sekanina noted that the collision of comet Shoemaker-Levy 9 with Jupiter in 1994 showed that the mass of a comet that enters the atmosphere of a planet, such as Earth or Jupiter, apparently needs to be more than 100 million tonnes in order to trigger a powerful explosion at the end of its journey. Smaller comets are likely to dissipate their mass in atmospheric flight and would end up with no appreciable mass at low altitudes. By contrast, the study of a number of fireballs pointed out that initial masses as small as 10 kilograms exhibited terminal flares. Various studies of the interrelation between a fireball's altitude, its pre-explosion speed, and the aerodynamic pressure at which it explodes corroborate the Tunguska fireball's pre-explosion mass of one million tonnes.

The question of a cometary hypothesis, Sekanina implied, 'now becomes mute'. He also noted that the Tunguska fireball was dwarfed by the comet Shoemaker-Levy 9 when measured by the amount of released energy. 'On the other hand, the Tunguska event directly involves the issue of a threat to our civilization,' he said.

Sekanina concluded his analysis by summarising eight points that made his asteroid theory broad-based and not based merely on eyewitness reports, as claimed by his critics:

- Its mid-air explosion is similar to the terminal flares of fireballs observed photographically.
- The pressure at the point of explosion, estimated at about 200 times normal atmospheric pressure, is consistent with a value that is expected from similar fireballs; if the object were a comet, the pressure would be about 2000 times atmospheric pressure, entirely

out of a plausible range for a fragile comet.
- The pre-explosion velocity of 36,000 km/h is the same as determined by seismic observations and a laboratory simulation of the uprooted forest.
- This velocity range rules out a comet-like orbit.
- Existing limited evidence on the event is inconsistent with a fragmentation pattern typical for comets.
- A comet of such a magnitude would have been extremely rare, perhaps 10 to 100 times more so than an asteroid of the same explosion energy.
- The limited evidence on the object's orbit is consistent with the orbits of Earth-crossing asteroids, but not with the orbits of short-period comets.
- This orbital information is particularly unfavourable to the hypothesis that associates the object with Encke's comet.

In 2001 a team of Italian scientists looked at the Tunguska object from a different angle, based on an idea of the late Paolo Farinella (1953-2000). Using data from detailed analysis of all the available scientific literature, including unpublished eye-witness accounts that have never been translated from the Russian, and the survey of directions of more than 60,000 flattened trees, the Italians plotted a series of possible orbits for the object. Of the 886 valid orbits that they calculated, 83 per cent of them were asteroid orbits, with only 17 per cent being orbits that are associated with comets.

This overwhelming data showed that the Tunguska object was indeed an asteroid. But if it was an asteroid why did it break up completely? According to one of the team, Luigi Foschini of the University of Trieste: 'Possibly because the object was like asteroid Mathilde, which was photographed by the passing Near-Shoemaker spaceprobe in 1997. Mathilde is a rubble pile with a density very close to that of water. This would mean it could explode and fragment in the atmosphere with only the shock wave reaching the ground.'

In 2005 Sekanina reflected upon his 1983 ground-breaking conclusion that the Tunguska object was an asteroid:

'I have not changed my views on the subject in the least, still considering Tunguska's asteroidal nature as virtually certain ... My involvement with the subject was in fact quite peculiar. I would not have begun my research on this subject in the early 1980s, if it were not for the fact that the papers published through the 1970s—all strongly pro-comet—had attracted my attention. I am a comet physicist and I felt that if it was a piece of a comet, one should be able to learn something on the properties of cometary nuclei from the event. Thus, I began to study the topic believing that this was indeed a comet: otherwise, I would have never got involved and I surely cannot be accused of initiating my research on this subject with a bias against Tunguska's cometary origin.

'My conclusion was simply the result of my findings: the more work I did, the more obvious it became to me that this was not a comet (just take the huge dynamic pressures to withstand!). By the time I was convinced completely, I invested so much time that I felt it would be a shame to drop the subject, even though it became essentially irrelevant to my own scientific interests (I never found asteroids to be much fun). And this was the sole reason behind my writing the 1983 paper. The rest is history.'

He adds: 'Of course, the cometary-origin hypothesis still has its old proponents such as Vitalii Bronshten, but no new admirers.' And, of course, the major contribution of Russian scientists in the development of the comet theory is well recorded in the history of the enigma known as Tunguska.

Smoking gun

In 1978 Ukrainian scientist Mykola Kovalyukh collected some fragments from a peat bog close to the epicentre of the Tunguska blast. Analysis of these samples showed that they contained a form of carbon, which has a crystal structure somewhere between diamond and graphite. Such arrangements are formed under extreme heat and pressure. However, the samples showed less of the metal iridium than is typically found in objects from the outer space. So Kovalyukh concluded that the samples came from terrestrial rocks altered by heat and pressure.

Twenty-five years later, Victor Kvasnytsya of the National Academy of Ukraine and his international colleagues, studied the sample using new high-resolution analytical and imaging techniques. In the journal *Planetary and Space Science* (vol. 80, 2013) they claim that the patterning and combination of minerals in the samples are very similar to that in other iron-rich, stony asteroids.

Is it indeed the smoking gun that could settle the enduing Tunguska mystery? It will take some convincing because Tunguska attracts so much speculation.

Many geoscientists have dismissed Kvasnytsya's claim as it does not yet have the conclusive proof. 'We get a lot of meteorite material raining down on us all the time,' Phil Bland, a meteorite expert at Curtin University in Perth, Australia, told *Nature*. Without samples of adjacent peat layers for comparison, 'it's hard to be 100% sure that you're not looking at that background'.

Target Earth

Are we going to be hit by a Tunguska-like asteroid again? Astronomers suggest that the average frequency of impacts of this size (average width 75 m) is one in 1,000 years. These asteroids explode in the lower atmosphere but release enough destructive energy to wipe out a large city.

The rate of impact decreases with the increase in size of the asteroid. The average interval between impacts of giant asteroids (average width 16 km) is 100 million years. An impact of this magnitude could destroy an entire continent and trigger mass extinction of advanced forms of life. Such an impact is believed to have wiped out the dinosaurs around 65 million years ago.

Do we really want to win an asteroid lottery? The probability of randomly picking six numbers in a lottery of 45 numbers is one in 4 million. The probability increases to one in 14 million if you have to randomly pick six numbers from 49 numbers, and to one in 19 million if you have to pick from 51 numbers. The probability of an asteroid impact, small or big, is one in 20,000, the same probability as for a passenger aircraft crash. From these odds it appears that the proverbial man in the street, if he is not run over by the proverbial bus (probability one in 100), will witness an asteroid impact long before he wins the big lotto. Why bother to buy a lottery ticket today?

Should we dismiss these risk probabilities as lies, damned lies and statistics, or lose sleep over the asteroid threat? What do the experts say?

'We do not know whether a large dangerous asteroid with our name on it is destined to hit [this] century,' says Britain's Astronomer Royal, Martin Rees. 'The risk isn't large enough to keep anyone awake at night, but it isn't completely negligible either.'

The American planetary scientist Tom Gehrels has a similar opinion: 'The chances of a celestial body colliding with Earth are small, but the consequences would be catastrophic.'

'Nobody believed Chicken Little when he said that the sky was falling. But occasionally the sky does fall, and with horrendous effects.' That's what Eugene Shoemaker, who was chiefly responsible for alerting the world to the dangers of asteroid and comet impacts, said in 1994.

'Doubters can ask the dinosaurs for their opinion,' advised a *Scientific American* editorial in November 2003.

Besides asking dinosaurs' expert opinion, what else could scientists do to stop a rogue asteroid from crashing into Earth? Once scientists have discovered an asteroid, they can calculate whether it is headed our way. The plan to destroy it will depend upon how far away it is. Here are some of the plans to save us from the scenarios dramatised in the 1998 movies *Armageddon* and *Deep Impact*:

Nudge 'em. If a threatening asteroid is spotted one year before the expected collision, it would be possible to nudge it with conventional-chemical fuel missiles. A small change in the asteroid's speed, on the order of 36m per hour, will deflect the asteroid 6,000 km—the radius of the target Earth—in one year.

Push 'em. Instead of a gentle nudge, some scientists prefer a stronger push by a nuclear-powered spacecraft that expels jets of plasma. The unmanned 'space tug' would rendezvous with the killer asteroid, attach to its surface and slowly push it so that it misses Earth.

Crush 'em. This ingenious plan suggests crushing the asteroid by placing a three-dimensional lattice made of millions of small tungsten balls in the path of the speeding offender. The collision would create enough heat to turn the asteroid into small, harmless rocks. The lattice could be launched into space by a rocket.

Cover 'em. Another ingenious plan is to wrap the rogue asteroid with a shiny plastic sheet like a giant potato in aluminium foil by sending a solar sail spacecraft that collapses around the asteroid and shrink-wraps it. An asteroid radiates heat into space after the Sun warms its surface, which imparts a tiny momentum to the asteroid, slightly shifting its orbit. A shrink-wrap or a spray of white chalk or black carbon powder across an asteroid's surface would change its reflectivity and, hence, the heat transfer from its surface. This heat transfer would change the momentum, which could be enough to change its path. After gift-wrapping it would be years before the asteroid changed its course.

Nuke 'em. If there is not enough time to prepare, some scientists suggest destroying the asteroid with a nuclear bomb. The idea is that the nuclear blast would melt material off the asteroid's surface, giving it a kick in the opposite direction. This is not really a clever way of killing an asteroid: if the bomb detonates too close to the asteroid, it may explode it, creating millions of tonnes of radioactive dust and rubble. If the rubble rains on Earth, we may have to join the dinosaurs anyway. Neutron bombs—the bombs that kill people but leave buildings undamaged—offer a better alternative. Once the high-energy neutrons hit the asteroid, they would heat its surface. The vaporised material would deflect the asteroid from the collision path.

Burn 'em. This plan involves placing a huge aluminium concave mirror fairly close to the asteroid. The mirror would focus a beam of light on a small spot on the asteroid. The heat would vaporise a small part of the asteroid which would shoot into space, pushing the asteroid in the opposite direction. An 800-m-wide mirror could deflect an asteroid 3 km in diameter.

Dig 'em. This simple but technically difficult plan requires placing a robot on the surface of the asteroid. The robot would dig rocks from the asteroid and hurl them into space, causing it to accelerate slowly in the opposite direction.

On the matter of the doomsday rocks, Tom Gehrels has the last word:

> Comets and asteroids remind me of Shiva, the Hindu deity who destroys and re-creates. These celestial bodies allowed life to be born, but they also killed our predecessors, the dinosaurs. Now for the first time, Earth's inhabitants have acquired the ability to envision their own extinction—and the power to stop this cycle of destruction and creation.

Chapter 5
Traceless Tunguska

Since Kulik's first expedition in 1927, there have been numerous Russian and international scientific expeditions to the Tunguska explosion site. Still, no impact craters or substantial meteorite remnants have yet been found. Besides a flattened forest, did the fireball leave other fingerprints?

Those who believe that the Tunguska object was either a comet or an asteroid also consider that the mass of the object vaporised into microscopic debris when it exploded at some height above the ground. Such a theory would also account for the absence of any impact craters. After the explosion, some of the microscopic debris drifted westward from Tunguska, but the rest condensed into microscopic globules that rained on to the Tunguska taiga. Many scientists have tried to interpret these fingerprints.

Early attempts

Scientists' attempts to track marks and traces of Tunguska began in 1957 when the Russian mineralogist A.A. Yavnel analysed soil samples collected by Kulik in 1929 from the Tunguska site, but these samples were later proved to be of terrestrial origin. Later, mineralogist O.A. Kirova, who was a member of Kirill P. Florenskiy's 1958 expedition, recovered both magnetite globules and various forms of silicate globules from samples collected from the region of the fall.

The magnetite globules were either shiny or dull, and their sizes ranged from 5 to 450 micrometres. The silicate globules varied in appearance from opaque to completely transparent and their sizes varied from 20 to 350 micrometres. The opaque silicate globules also contained large numbers of gas bubbles and traces of iron oxide. Both types of globule are characteristic of the particles produced by the destruction of a meteorite in the atmosphere.

Florenskiy's next expedition in 1961-62 concentrated on the study of the distribution of meteorite material in the soil. This was the first expedition to use a helicopter to deploy individual groups and transport heavy equipment. The expedition collected thousands of tonnes of soil samples at regular distances of 10, 20, 40, 60 and 80 km from the epicentre.

The analysis of these samples showed brilliant magnetic (magnetite) and glassy (silicate) globules, less than 0.1 mm in size. When a map of the distribution of these globules was drawn up, it showed that the globules occurred over a fairly well-defined ellipse, with high concentrations between 60 and 80 km to the northwest of the epicentre. This pattern of distribution was probably the result of the wind direction of the day, which was from southeast to northwest. The appearance of the globules showed that they were formed in the atmosphere as molten matter condensed. Florenskiy was convinced that most of the globules were the remnants of the Tunguska comet that exploded in mid-air.

In the early 1970s the Soviet scientist G.I. Petrov contended that as the Tunguska meteorite moved through the atmosphere, it rapidly evaporated and, when a large amount of vapour had amassed in front of the travelling body, it exploded and scattered in the atmosphere. This scattered matter then condensed into microscopic balls that settled on the vast Tunguska forest. Sphagnum bogs, which receive mineral nutrients only from the air, are most likely to assimilate such 'meteorite ball.'

As a result, Soviet scientists analysed the peat layers of the sphagnum bogs, which add a distinct layer of peat annually. They discovered fused silicate globules, up to 0.8 mm in diameter. The concentration of globules was substantially higher in the 1908 layer than in layers before and after the catastrophe. These globules were composed of rare-earth and heavy elements—the elements present in extraterrestrial material.

In his last paper published in English before his death in 2001, Academician Nikolai Vasilyev expressed doubts about the origin of the globules, as similar globules could also be formed during peat burning. 'The presence of small quantities of meteorite dust cannot be doubted,' he wrote in *Planetary and Space Science*, 'but the problem of their connection to Tunguska remains open.'

Another attempt to study peat was made in the late 1970s by Emil Sobotovich and his colleagues at the Institute of Geochemistry and Mineral Physics in Kyiv. After six years of investigation, they concluded that the Tunguska blast was caused by a 4,000-tonne stony meteorite that exploded before impact, scattering particles over a vast region. The researchers presented convincing evidence of their claim: a large number of tiny diamonds strewn over the Tunguska region. The team incinerated peat collected from the region of the fall in high-temperature ovens. In the ashes, they found many irregularly shaped and extremely hard black grains. Further laboratory examination showed them to be diamonds.

Diamonds are formed under extraordinarily high pressures in volcanic pipes known as kimberlites. Since no such pipes have been found in the Tunguska area, the Kyiv researchers concluded that the Tunguska diamonds were formed far from Earth (see the section 'The escapade of a gas' in Chapter Eight which contradicts this claim). Such diamonds are already known to exist in uralites, a class of meteorites, which presumably have been involved in deep-space impacts. These meteorites may originate ultimately from comets.

Other researchers have suggested that the Tunguska object was a chunk of comet Encke, a periodic visitor to our skies. To support their claim that the tiny diamonds arrived in a meteorite, Kyiv researchers cited high levels of radioactive carbon-14 in the peat. They said that such high levels of carbon-14 are found in meteorites that have been subjected to prolonged bombardment by cosmic rays in space.

Tunguska imprints in Antarctica

Globules collected by Florenskiy's 1961-62 expedition were enriched in iridium, a silvery metal that is abundant in extraterrestrial bodies but rare on Earth and contained other evidence of extraterrestrial origin. Traces of the Tunguska fireball were not only present in the Tunguska soil; they have also been discovered in an Antarctic ice core.

This startling revelation—made in 1983 by Ramachandran Ganapathy, an American expert on extraterrestrial materials—provided fresh insights into the nature of the Tunguska object.

Ganapathy examined eight globules by a nuclear analytical technique to find the answer to three questions:

- Are the globules truly extraterrestrial?
- Are all the globules related to each other as would be expected if they originated in the Tunguska explosion?
- Can these globules be distinguished from the remnants of iron meteorites?

All eight globules contained iridium, nickel, cobalt, gold, chromium, antimony and iron. Their iron content ranged from 76 to 81 per cent, and they were all enriched in iridium, a reliable indicator of extraterrestrial matter. One of the globules had a whopping 56,900 parts per billion iridium (rocks on Earth's surface contain only about 0.3 parts per billion iridium; a meteorite may contain as much as 500 parts per billion). The concentration of nickel and cobalt, two elements always found with iridium in the extraterrestrial matter, was also present in cosmic proportions. 'There is no question about it,' Ganapathy declared in *Nature*, 'the spheres are extraterrestrials.'

The identical ratio of iridium and nickel in each globule also proved that they all came from the same extraterrestrial body. The result also disproved the notion that the globules were from meteors falling continuously on Earth. The high abundance of chromium in some globules, however, indicated that the object was not an iron meteorite.

Ganapathy also reasoned that the eyewitness accounts of the Tunguska explosion, together with his findings, demonstrated that the explosion in the atmosphere vaporised the Tunguska object was. The question of whether the debris from the event could have reached the stratosphere, and as a consequence been distributed globally, prompted him to search for debris in the Antarctic ice. Old ice and air bubbles trapped in it are precious scientific tools. They can illuminate the past, help answer questions about the present, and assist prediction about the future.

The average rate of accumulation of ice—about 7 cm per year at the South Pole—is a reasonably good chronometer for measuring time with depth. For his study, Ganapathy selected a 101-m ice core drilled in 1974. A sample from a depth of 10.15 to 11.07 m, which corresponded to 1908, contained four times more iridium than found in earlier years. Ice was similarly enriched in iridium during 1909 and the next few years. Then the iridium concentration dropped back to normal levels. This was the strongest proof that particles from the Tunguska fireball were scattered around the world by atmospheric currents. 'Because this iridium could only be deposited here by means of stratospheric fallout, it should be presented worldwide,' he said.

The amount of iridium deposited in Antarctica could be used as a clue to calculate the total amount of atmospheric fallout from the event. The result: 7 million tonnes of debris. Ganapathy estimated that the object that exploded over Tunguska was a 7-million-tonne, 160-m-diameter monster. He warned that his estimate did not make a distinction between comets and asteroids. However, he said that the object 'may well have been a stony asteroid'.

A year after Ganapathy's observations, Polish scientist Marek Zbik examined 100 black magnetic globules from the Tunguska area. The globules varied both in size (from 7 to 350 micrometres) and shape (spherical, droplet-like, some even broken and damaged). Many of these globules turned out to be of terrestrial origin. The ratio of iridium and nickel in the remaining globules was very close to that observed by Ganapathy, proving them to be of extraterrestrial origin. Zbik's analysis did not establish a definitive link between the globules and Tunguska.

The testimony of trees

Carbon atoms come in eight varieties, known as isotopes. Carbon-12 (six protons and six neutrons in the nucleus) is the most common isotope. High-energy neutrons which continuously bombard Earth convert ordinary carbon-12 into radioactive carbon-14 (six protons and 8 six neutrons). Living things go on absorbing carbon-12 and carbon-14 until the time of their death. In the case of trees, carbon-14 is recorded in annual growth rings, which also give the age of trees.

It has been suggested that a comet caused the Tunguska explosion. Therefore, the hydrogen contained within it would have been compressed and heated as the comet passed through the atmosphere. Some of the hydrogen might have fused into helium, triggering a nuclear explosion that would generate high-energy neutrons and consequently carbon-14 in the atmosphere.

Many scientists have measured carbon-14 in Tunguska tree rings, corresponding to pre- and post-Tunguska years. Examination of tree rings formed in 1908 shows a rise in carbon-14, but not enough to support the idea of annihilation caused by a nuclear explosion. The surge in carbon-14 is attributed to the solar cycle, in which sunspot numbers rise or fall over a typical period of eleven years. It has also been suggested that the burn-up of the Tunguska object in the atmosphere would have produced a temperature of a few million degrees, too low for nuclear reactions but high enough to produce carbon-14.

A ground-breaking experimental approach by a team of Italian scientists from the University of Bologna, headed by Giuseppe Longo, has uncovered new remnants of the Tunguska fireball. One of the team members, Menotti Galli, was on the 1989 Tunguska expedition, the first post-Cold War expedition open to international scientists. Galli, a physicist, is an expert on phenomena associated with cosmic radiation, including carbon-14.

During the expedition, Galli realised that the only witnesses to the 1908 blast still alive were the surviving trees. But their testimony was hidden in the resin formed around broken branches after the blast. Like amber, this resin could have acted as a trap for particles present in the atmosphere, including extraterrestrial particles from the fireball. The resin would harden and form a protective coating around the branch. Eventually, the resin would become enclosed within the growing branch. What Galli needed was to examine trees in the blast area. Their annual growth rings would point him to the 1908 sections, and if the fireball had showered any particles on the forest, they could still be intact in those sections of the trees. Bingo!

To collect the samples for examination, Galli and his colleagues, Longo, a nuclear physicist, and Romano Serra, an astronomer, attended the 1991 expedition. 'The Italians, accustomed to sipping espressos under Bologna's endless porticoes, found themselves slaking their thirst with brown swamp water laced with mosquito larvae,' says Richard Stone in *Discover* magazine (September 1996), describing the 'ten difficult days' spent by the Italian scientists in Tunguska.

With or without espressos, they still managed to collect resin deposited between 1885 and 1930 on 14 branches of seven Siberian spruce trees abundant in resin. The trees were situated in different directions within a radius of 8 km from the blast's epicentre. For comparison, they also collected resin from six branches of a tree growing at about 1,100 km from the Tunguska site, and the roots of a tree uprooted by the blast.

Back in Bologna, the researchers used a scanning electron microscope to examine their samples. In all, they recovered 5,854 particles from the Tunguska branches and 1,183 particles from the two control trees. Their examination of these microscopic particles showed an anomalously high abundance of iron, calcium, aluminium, silicon, gold, copper, titanium, nickel and other elements. Some of these elements are commonly associated with normal-density stony asteroids. This abundance peaked around 1908. Another interesting observation was that the smooth texture and spherical shape of particles from the Tunguska branches showed evidence of heating and melting. 'The blast wave would not have melted particles in the ground, where the conductivity was low,' said Longo. 'That means the melted particles came directly from the cosmic body.'

The sample of trees that survived the 1908 blast taken from the North and South sides of Lake Cheko show that they had tight rings before 1908, meaning the trees grew very slowly due to competition with other trees and were growing densely together. After 1910, the trees show much wider rings, indicating there was less competition with other trees, more sunlight and nutrients.

Surviving trees were smaller during the 1908 blast suggesting that they were bent over or twisted during the impact. All of the large trees, on the other hand, were uprooted. This typical of what happens during a cyclone.

Vasilyev pointed out in 1998 that the elements discovered by Longo's team in tree resin were similar to those found by Russian scientists in peat layers. This effect is most probably connected to the Tunguska body. Still, for the final identification of the particles found in resin as the Tunguska matter, some additional corroboration is necessary, because a large volcanic eruption in Russia on 28 March 1907 had produced a significant dust veil over the Northern Hemisphere for more than a year. 'There is no direct evidence that these materials have anything to do with the Tunguska body,' he said. 'On the contrary, there is good reason to believe we are dealing with fluctuations in the background fall of space dust.'

In the same year, Vladimir Alekseev of the Troitsk Institute for Innovation and Fusion Research in Moscow also expressed doubts on the methods used by Longo's team because of the presence of background particles which could come from volcanic eruptions. He came up with a different approach for hearing the testimony of trees: examining particles more energetic than those of the background. For his study, Alekseev selected a standing larch tree that had survived the Tunguska catastrophe. The tree, located near the epicentre, had a 10-cm vertical split in the stem. The split, according to Alekseev, could have been caused by shock waves which, coming from above, exerted a force on the growing tree. He took a wood sample from the split.

After removing resin from the sample, when Alekseev examined it with a high-powered microscope, he noticed numerous solid particles up to 50 micrometres in size in the dense wood of the 1908 growth ring. The particles could be divided into four groups: metallic particles with jagged edges; spherical silicate particles; whitish particles; and black graphite-like particles. They had energies high enough to penetrate the dense wood, and Alekseev was convinced that they were the remnants of the Tunguska body.

Based on the information obtained from the study of these particles, Alekseev proposed the following scenario for the Tunguska blast: multiple explosions finished the flight of the body and, therefore, the solid remnants are small particles. The numerous explosions could be responsible for the gunfire-like sounds repeatedly heard by the eyewitnesses of the event. There was a possibility of thermonuclear reaction on the surface of the body at the final stage of its journey in the atmosphere. Like some cosmic bodies, this body was probably enriched with deuterium (hydrogen 2). This deuterium could start a thermonuclear reaction which would convert deuterium to tritium (hydrogen 3). 'As tritium is involved in biological processes and it is radioactive, it can have genetic effects,' Alekseev concluded.

New evidence from Lake Cheko

In 1960, a Russian scientist, V.A. Koshelev, speculated that Lake Cheko might be an impact crater. This suggestion was rejected by Kirill P. Florenskiy, a main proponent of the comet theory in the 1960s, after having found loose, thick sediments below the surface of the lake. The presence of these sediments suggested that though the lake was formed by the impact of a cosmic body, it was definitely formed before the 1908 catastrophe.

In the 1990s, Giuseppe Longo and his colleagues, stumbled on the obscure papers of Koshelev and Florenskiy that reported their discovery of Lake Cheko and its origin. In recent years the three Italians have spent considerable time and effort to study Lake Cheko. This 500-m wide and 47-m deep lake is 8 km from the centre of the 1908 explosion

During the second Italian expedition to Tunguska in 1999, the scientists not only continued their search for microparticles preserved in tree resins, but they also looked for other remnants in the sediments at the bottom of Lake Cheko. The group, which included Longo who had also attended the 1991 expedition, used an inflatable catamaran for the geological survey and the coring operations.

This work had two objectives: to check whether the lake is an impact crater of the 1908 event; and to detect mineralogical, chemical and biological evidence of the nature of the Tunguska fireball. Their experimental study showed that though the impact of a cosmic body formed the lake, it was formed before the 1908 catastrophe. The core samples collected from the sediment have not yet conclusively been shown to be linked with the fireball.

After their 2007 on site studies of the lake using a range of technologies such acoustic-echo sounder, magnetometer and ground-penetrating radar (GPR), and analysing local geophysical and geological records, the Italians conclude that Lake Cheko is young, not older than the Tunguska blast. The funnel-like structure of this 50-m-deep lake is very different from that of common Siberian lakes. If the lake were thousands of years old, it would probably have a flat bottom, the result of fine sediments gradually filling up.

Most computer models of the blast suggest that the Tunguska body disintegrated and vaporised 5-10 km above the ground. However, these models do not exclude the probability that a large fragment hit the ground at a relatively low angle of 45 degrees forming a crater, which was finally filled by water from the nearby Kimchu river, forming Lake Cheko.

Total ablation

Vladimir Svetsov of the Institute for Dynamics of Geospheres in Moscow removed a layer of mystery from the Tunguska event when in 1996 he showed that the entire mass of the Tunguska body vaporised before it could reach the ground. Ablation—the loss of material from a cosmic body through evaporation or melting caused by friction within the atmosphere—of the Tunguska debris was total.

According to Svetsov, when a Tunguska-sized body enters deep into the atmosphere, it is broken into a large number of fragments, the maximum size of which is 10 cm. As the body decelerates, these fragments are separated from each other. Mathematical simulations show that stony fragments fully ablate either inside or outside the fireball due to high temperatures. All vaporised material does not reach the ground; it moves upwards in the atmosphere. Although Svetsov admitted that fragmentation processes were complex and certainly needed further investigation, he declared his scenario 'quite plausible'.

As for the Tunguska body, Svetsov said that it would have been heated to 15,000°C. This temperature was high enough to create an explosion 'quite comparable with that of a nuclear explosion'. Upon explosion, the body broke into a vast number of fragments, typically 1 to 3 cm across but no larger than 10 cm. But the temperature was high enough to melt these fragments until nothing remained. Some of the microscopic debris condensed in the atmosphere and then was scattered over the Tunguska forest. 'The apparent absence of solid debris is, therefore, to be expected following the atmospheric fragmentation of a large stony asteroid,' Svetsov concluded.

He also said that similar to the 1994 impact of comet Shoemaker-Levy 9 on Jupiter, Tunguska debris was probably also widely scattered because of the turbulent wake of the asteroid. Svetsov also suggested that the microscopic particles recovered from the resin of trees by the Italian researchers 'might be recondensed material precipitated in the general vicinity of the impact site'.

Svetsov did leave some hope for Tunguska meteorite hunters: if some larger fragments accidentally gained significant speeds at altitudes between 15 and 20 km, sizeable remnants could reach the ground. But they would have fallen to the ground at a distance of 5 to 10 km southeast from the blast's epicentre.

Svetsov's calculations assumed that the Tunguska fireball was a 15-megaton asteroid-like object that hit Earth with a speed of 54,000 km/h and at an angle of 45 degrees. What if the object was a comet? This analysis would then make no sense. NASA scientist Kevin Zahnle provides an argument that comet partisans would find hard to demolish. His argument goes something like this: the relatively rare iron meteorites almost always produced small impact craters on Earth.

The 1.2-km-wide Meteor Crater in Arizona, for example, was produced by an iron body of mostly the same energy as the Tunguska explosion. The smallest known crater made by a stony meteorite is the 3.4-km New Quebec crater. This raises a problem for the comet theory: if comets with energies of 15 megatons can reach the higher atmosphere before exploding, then the much more numerous asteroids, which most astronomers agree will penetrate deeper, should be cratering the land every 1,000 years. 'If Tunguska was a comet, where are all the Meteor Craters made by rocks?' Zahnle asks.

Environmental effects

During his first expedition to Tunguska in 1927, Kulik noticed quite a rapid recovery of the forest after the catastrophe. As we have seen, he wrote in his diary, 'The young twenty-year-old forest growth has moved forward furiously, seeking sunshine and life'. Many Russian scientists have studied the accelerated growth of trees that survived the catastrophe. They have noticed that the effect does not coincide with the limits of the fire and destroyed forest and is observed not only in surviving trees but also in younger trees germinated after the catastrophe.

It has been suggested that this accelerated growth is the result of genetic mutation caused by a nuclear explosion. Longo's team examined the abundance of carbon-14 in the 1903-16 tree rings but found no traces of nuclear processes. This observation contradicted Alekseev's belief that radioactive processes were possible on the surface of the Tunguska body. But the Italian team suggested that the accelerated tree growth seemed to derive from such improved environmental conditions after the explosion as ash fertilisation by charred trees, decreased competition for light, and greater availability of minerals due to increased distance between trees.

A detailed analysis of the environmental effects of the Tunguska catastrophe was conducted by the American atmospheric scientist Richard Turco and his colleagues in the 1980s. Turco's analysis assumes that the Tunguska object was an icy comet nucleus, rich in water, ammonia, carbon dioxide and methane. As this object passed through the atmosphere, these substances contributed to the production of 30 million tonnes of nitrogen oxide.

After comparing nitrogen oxide generated in nuclear bombs, Turco concluded that the Tunguska event might be compared approximately to 'a large-scale 6,000-megaton nuclear "war" in terms of nitrogen oxide deposited in the stratosphere'. In addition to this massive injection of nitric oxide in the stratosphere, the object also added about 1.5 million tonnes of water. This water helped in the formation of noctilucent clouds which caused bright nights. But the nitric oxide contributed to a long and deadly effect. By a complex series of reactions, nitric oxide converts stratospheric ozone into oxygen. These reactions depleted the ozone layer that protects us from harmful ultraviolet rays.

The dust veil that hung over the stratosphere for years also contributed to climate changes. Turco said that about one million tonnes (Ganapathy's estimate: 7 million tonnes) of dust would be likely to decrease the average surface temperature by about 0.05°C. This resulted in an overall cooling of about 0.2 to 0.3°C in the Northern Hemisphere. Turco pointed out that the 1907 volcanic eruption in Russia might have initiated the cooling trend. Turco's team also studied weather records from the early 1900s and noted several other unusual weather conditions that appeared to begin around 1908 and lasted for several years:

- An increase (above a decreasing trend) in the mean surface temperature over North America in both January and July beginning in 1909-10.
- An increase in total Arctic ice between 1908 and 1911.
- A 50 per cent decrease from typical values in the number of tropical cyclones in the Atlantic and Caribbean Oceans.

Turco's team concluded its study by saying that 'this most impressive and consequential natural event' might have historical significance. Ozone depletion and climatic changes associated with large meteorites may have had a role in past events, such as the death of the dinosaurs 65 million years ago.

Sure, we now know about the environmental impacts of the Tunguska object and its remnants. Still, there remains the little matter of the identity of the object: was it a comet, an asteroid or something else?

Chapter 6

The Incredible Journey of a Black Hole

Mini black holes are hypothetical tiny black holes. Still, they provide such a neat and fitting explanation for the Tunguska event that this theory has become the part of the folklore now associated with the mysterious fireball. In 1973 the American theoretical physicists A.A. Jackson and Michael P. Ryan said that, since 'no crater and no meteorite material that can unambiguously be associated with the event have ever been found,' a mini black hole could explain the Tunguska event.

In an article in *Nature*, Jackson and Ryan suggested that, after passing through the atmosphere, the mini black hole would have entered Earth. Because of the rigidity of the rock, there would have been no underground shock wave. With its high velocity, the black hole would have passed straight through Earth in about fifteen minutes and exited through the North Atlantic, causing shock waves in the ocean and the atmosphere.

The mini black hole was much smaller than the full stop at the end of this sentence, but it had the mass of a giant asteroid and quite a strong gravitational field for some distance from the body. As it passed through the air, it became scorching hot, producing a deep blue trail of particles. Jackson and Ryan based their argument on the assumption that the damage caused by the Tunguska fireball was equivalent to a two-megaton nuclear explosion. They estimated that the total energy in the black hole's blast wave would be within the same range.

Meet a monstrous heart of darkness

An enormous cloud of gas and dust in the centre of a galaxy condenses to form a compact nuclear disc. As this disc condenses further, it becomes unstable and collapses under its gravity into a point of zero volume and indefinite density—a singularity. At this point, the laws of physics break down, and space has no meaning and time stops.

The mass of this newly created enigmatic entity is more than 10 billion Suns, yet it has a voracious appetite for more matter. It devours stars that come close to it. Matter falling into disappears forever. It's a black hole, a supermassive black hole.

The matter falling into the black hole forms a large disc of gas and dust around it. This disc, called the accretion disc, would also eventually fall back into the black hole. The boundary separating the singularity and the accretion disc is the event horizon, a kind of one-way spherical edge of the region of no escape.

Nothing—not even light—can escape the event horizon. The radius of the event horizon of this black hole is about five times that of our solar system.

As no light can escape the black hole, you cannot see it even with the most powerful telescope. When matter from the accretion disc approaches the event horizon, it spirals around in loops with hypervelocity. If it collides with anything else before reaching the event horizon, it releases enormous kinetic energy which converts into motions of atomic and subatomic particles and electromagnetic radiation. Formed before reaching the event horizon, the particles and photons travel outward brightening the galaxy around it. (Powerful telescopes can see the brilliance of the galaxy from which we can conclude that there is a supermassive black hole lurking at its centre. Of course, this doesn't provide absolute proof of the existence of the black hole.)

When you are close enough to this black hole, there is no place to hide. The light from your image would be wrapped around the event horizon and visible to someone on the opposite side. Black holes have this remarkable ability to bend light rays.

Throw a clock into this black hole—an impossible thing to do but imagine you could do to. It would slow down as it approaches the event horizon, and when it crosses the event horizon, it will freeze completely. There is no time to measure. It would also fade from view as gravity would stretch light from it, making it redder and redder until it appears from the view.

If you move closer, you would be sucked in. Its gravity would stretch your body into the shape of very long spaghetti. This process is known as spaghettification. The dead spaghetti then slams into the singularity of the black hole, crushing your remains to an infinite density.

As this black hole swallows up more matter, it swells. But you will never notice any change in it, except the increasing size of the radius of its event horizon.

The radius of a black hole is the radius of the event horizon surrounding it. This is called the Schwarzschild radius, after the German astronomer Karl Schwarzschild who in 1916 predicted the existence of a dense object into which other objects could fall, but out of which no objects could ever come (the term 'black hole' was first used in 1969 by the American physicist John Wheeler; before that they were known as 'collapsars' or 'frozen stars').

The Schwarzschild radius is roughly equal to three times the weight of the black hole (in solar masses). A black hole weighing as much as the Sun would have a radius of 3 km; one with the mass of Earth would have a radius of only 4.5 mm, and one with the mass of a small asteroid would be roughly the size of an atomic nucleus. A black hole's weird effects occur within 10 Schwarzschild radii of its centre. Beyond this somewhat limited distance, the only effect is through the black hole's normal gravitational pull. So, contrary to popular belief, a black hole is not like a cosmic vacuum cleaner that sucks in everything around it.

The eminent theoretical physicist Stephen Hawking, who has greatly advanced our knowledge of black holes, has said that 'black holes are not black after all: they glow like a hot body, and the smaller they are, the more they glow'. He proposed a mechanism by which black holes transform their mass into both radiation and particles that leave the hole.

The result is that black holes gradually evaporate. So, they do not last forever. The amount of radiation, now known as the Hawking radiation, escaping from a black hole is inversely proportional to the square of its mass; that is, the smaller the black hole, the shorter its life span. A primordial black hole (primordial black holes formed soon after the Big Bang) with the initial mass of Mount Everest (and the size of an atomic nucleus) would have a lifetime roughly equal to the age of the universe, that is, 13.8 billion years. But a black hole with the initial mass of Sun would vanish after about 100 million billion billion billion billion billion billion (1 with 62 zeros after it) years.

Mini black could have masses ranging from a speck to tens of thousands of times the Sun. They could have formed in the fraction of a second after the Big Bang.

The story of Big Bang

If you wish you can begin every story with the ages-old phrase 'once upon a time', but not this one. Time doesn't exist yet. Everything you can view has a perspective, but not the event about to happen. There is no space—no below, no above, no 'outside' in which to stand and watch the primeval explosion. There is no matter, no energy (therefore, no light). It's intensely dark, dark like darkness hiding darkness.

There is nothing except something infinitesimally small and unimaginably dense and —something called a singularity—that defies the laws of physics. And then this singularity explodes spontaneously. This explosion is the Big Bang. Matter, energy, time and space begin at this instant—and the laws of physics come into play (and Einstein can rejoice).

The Big Bang is not like a big bomb going off in a particular location in an otherwise empty space. Space doesn't exist. The Big Bang is an explosion of space itself. The explosion has no centre; it's happening everywhere. It's just like the inflation of a balloon on a scale which makes the words like gigantic and humongous look ridiculously puny.

After one billion trillion trillion trillionths of a second (10^{-43} second), the universe starts inflating. Space and time disentangle; earliest meaningful time begins. Matter and energy are interchangeable.

The universe cools as it blows outwards, and by one hundred billion trillion trillionths of a second (10^{-35} second) cosmic expansion creates a large, smooth patch of space filled with quarks, the building blocks of matter. The rate of expansion is now mind-boggling: imagine a pea growing to the size of the Milky Way in less time than it takes to blink.

After one hundred billion trillionth of a second (10^{-23} second), the universe is a superhot soup of electrons, quarks, photons and other particles. The soup's temperature is one thousand trillion trillion (or 10^{27}) degrees Celsius.

After 10 microseconds, quarks combine to form protons and neutrons; and after 10 milliseconds, protons and neutrons begin to form hydrogen and helium nuclei.

At about 380,000 years, the universe cools to approximately 3000°C and protons, neutrons and electrons combine to form other atoms. The newly born universe was opaque as there was no light, but the superhot soup now breaks up and photons, particles of light, stream across space in more or less straight lines. The universe is now transparent, and other forms of electromagnetic radiation are free to escape as gamma rays. As the universe continues to cool and expand, gamma rays change into X-rays, ultraviolet rays, visible light and then finally into microwaves.

After 100 million years, first stars and dwarf galaxies form from hydrogen gas that now permeates the universe. The first rays of starlight end the darkness of the universe.

After 300 million years, the dwarf galaxies start merging and building up bigger galaxies.

After 9 billion years, the Sun and Earth are born.

After 13.8 billion years. Because space is expanding, the universe, which is an expanse of flat space, has now a radius of more than 13.8 billion light-years. It has more than 200 billion galaxies. Microwaves fill the universe and can be detected anywhere. This leftover warmth from the primeval fireball, known as the cosmic microwave background, has a temperature of −270°C or 2.7 kelvin (3°C above the absolute zero).

The Tunguska black hole

When Jackson and Ryan proposed their impeccably scientific black hole theory, they commented that many attempts had been made to explain the Tunguska event, 'ranging from the prosaic to the bizarre'. They then suggested that 'a black hole of substellar mass such as those that have been postulated by Hawking could explain many of the mysteries associated with the event.' Their explanation has never been called 'prosaic' or 'bizarre,' but it has indeed been described as 'imaginative and intriguing' by some scientists.

How can a small black hole explain the Tunguska event? Jackson and Ryan's case was based on three main arguments:

High speed. The researchers assumed that the black hole had the mass of a large asteroid (about 100,000 billion to 10,000,000 billion tonnes), but its geometrical radius could be measured in micrometres. However, its gravitational field could be quite strong for some distance from the body. They also assumed that the black hole's escape velocity—the minimum speed an object must have to free itself from the gravitational pull of a planet or a star—was slightly greater than Earth's escape velocity, which is about 40,000 km/h. They calculated that, if the black hole began in interstellar space with zero speed and fell freely to Earth's orbit, its speed relative to Earth would be between 36,000 and 360,000 km/h. Thus, the black hole would travel through the last 30 km of the atmosphere in about one second.

Bright blue 'tube'. The air around the passing black hole would heat to between 10,000 and 100,000°C. So, most of the radiation from the shock front would be ultraviolet rays. The accompanying plasma column would, therefore, appear blue. 'These results agree well with eyewitness reports of the event and with measurements of the pattern of throw-down of trees at the site,' Jackson and Ryan said.

No crater. 'Since the black hole would leave no crater or material residue, it explains the mystery of the Tunguska event,' they said. 'It would enter Earth, and the rigidity of the rock would allow no underground shock wave. Because of its high velocity and because it loses only a fraction of its energy in passing through Earth, the black hole should very nearly follow a straight line through Earth, entering at 30 degrees to the horizon and leaving through the North Atlantic'. At the exit point, there would be another shock wave and disturbance of the sea surface. Jackson and Ryan suggested that oceanographic and shipping records could be studied to see if any surface or underwater disturbances were observed.

Scientists have found, forgive the pun, many holes in the black hole theory. 'The black hole would have shot straight through Earth but unfortunately for the theory (although fortunately for us) the exit point, latitude 40 degrees, 50 minutes north, longitude 30 degrees, 40 minutes west, in the mid-Atlantic was not marked by an equally severe shock and blast wave,' commented the British astronomer David Hughes in *Nature*.

Gerald Wick and John Isaacs of the Scripps Institution of Oceanography in California also wrote in *Nature*: 'Unfortunately, this miniature, hypothetical object cannot account for all the important phenomena known to accompany the event'. Their main argument centred around the small magnetic globules with high nickel content found in the Tunguska region. High nickel content confirms that the globules are of extraterrestrial origin, but it does not necessarily ensure that they originated in the Tunguska blast.

'Most soil samples collected at random over the globe will contain similar cosmic dust,' Wick and Isaacs said. 'The spatial pattern of the globules collected in the Tungus, however, shows that the cosmic dust likely originated from a massive meteorite body of dimensions vastly greater than a few angstroms.' However, they agreed that their discussion did not preclude the possibility that a black hole comprised the nucleus of the comet, or that black holes frequently might be the agents condensing the materials in such bodies.

William Beasley and Brian Tinsley of the University of Texas at Dallas claimed in *Nature* that several lines of evidence rendered the black hole theory extremely unlikely:

First, many characteristics of the event indicated that the main part of the energy went into an explosion in the air. These characteristics include trees scattered on the ground, without branches or bark, in the direction opposite to the centre of the fall; an intense fire that seared trees; and in ravines, partially protected trees that remained standing, but many with their tops broken. A typical meteorite buries itself below the surface and then dissipates its energy in an underground explosion. The Meteor Crater in Arizona was produced in this way. 'No significant excavation was caused by the Tunguska explosion,' remarked Beasley and Tinsley.

Second, a small cometary nucleus, consisting of a mass of frozen gases mixed with nickel-iron and silicate particles, would have a low degree of cohesion and would fragment in the air and dissipate most of its kinetic energy before it reached the surface. A small black hole could also produce a similar air blast. The black hole would, however, have passed through Earth in 10 to 15 minutes and caused a similar explosion at the point of exit, which would have occurred in the North Atlantic.

Third, about five hours after the Tunguska blast, six microbarographs in England recorded sound waves from the explosion. The approximate distance from the point of impact to the centre of the microbarograph stations is 5,720 km, so the average speed of the waves was about 1,150 km/h, which is about the usual value for this type of wave. The recorded waves were travelling from Siberia, and not from the North Atlantic. Sound waves from the site of the suggested exit explosion should, however, have arrived in England about three hours before the arrival of the Siberian wave. The Beasley and Tinsley stressed that they had examined copies of the English microbarograph records but had been unable to find any sign of waves from the North Atlantic exit point.

Fourth, a thick dust train along the path of the fireball immediately after its passage was noted by eyewitnesses. This observation is consistent with the deposition of the material in the atmosphere, rather than the loss of air into a black hole.

Fifth, exceptionally bright nights in Siberia and Europe imply that extraterrestrial material was deposited in the upper atmosphere simultaneously with the impact. 'The wide area of atmospheric deposition is comparable to the dimensions of a cometary tail,' Beasley and Tinsley said, 'and is not compatible with the idea of slow transport of dust vertically and horizontally from a ground-level explosion'. This deposition of dust in the upper atmosphere could give rise to noctilucent clouds, which could account for the bright nights.

'All the evidence favours the idea that the impact which caused the Tunguska catastrophe involved a body with characteristics like a cometary nucleus, rather than a black hole,' Beasley and Tinsley concluded.

A further challenge to the black hole theory came from the American scientists Jack Burns, George Greenstein and Kenneth Verosub. In the *Monthly Notices of the Royal Astronomical Society*, they discussed inconsistencies in the predicted and observed thermal changes of soil and rock and seismic activity associated with the event. 'The point of entry of the black hole into Earth should be marked by a patch of melted and resolidified rock of diameter half to 4 km, overlain by fused soil of comparable extent,' they said. 'As the hole entered the soil, it would have vaporised the water, oxidised the organic matter and fused the residual material such as quartz, feldspar and mica ... the point of impact should, therefore, be marked by a depression'. The Southern Swamp, or Kulik's Great Cauldron, is a depression. Still, they pointed out that 'this depression may predate the Tunguska Event and is not inconsistent with other explanations.'

Burns, Greenstein and Verosub's calculations showed that Jackson and Ryan's black hole would release seismic energy equivalent to one million to 100 million megatons of TNT in Earth. In contrast, the largest earthquake ever recorded (magnitude 8.3) released the equivalent of only 50 megatons. 'The absence of enormous seismic activity associated with the Tungus event, therefore, precludes its interpretation as a small black hole,' they declared.

A postscript to Jackson and Ryan's popular theory appeared in Rupert Furneaux's book *The Tungus Event* (1977). Upon learning that no exit pulse had been found on the English microbarographs described in Chapter 1, the scientists were disappointed at the rejection of their theory: 'It begins to seem that the Tunguska event is more bizarre than any explanation put forward to date.'

Many decades have passed since the publication of Jackson and Ryan's black hole theory. We do now have a better understanding of black holes. Does this new knowledge support their theory? No scientific theory prohibits a wandering black hole striking Earth, but the question is: did a mini black hole pass through Earth on the morning of 30 June 1908?

Chapter 7

The Matter in Question

The idea of the atom may have originated in Babylon or Egypt or even in India, but the story of matter started in 5th-century BC Greece with Leucippus and his pupil Democritus. They taught that matter was composed of space and an infinite number of tiny, indestructible particles called *atomos* or atoms. But Aristotle and other Greek philosophers preferred their 'elements'—earth, air, fire and water—out of which the whole world was created, and Democritus' idea was lost for two millenniums. It was recovered and expanded in 1808 by John Dalton, a Quaker schoolmaster from Manchester, into his atomic theory.

The first real picture of the atom emerged in 1897 when the British physicist J.J. Thomson suggested that atoms are like a Christmas pudding, in which negatively charged electron 'raisins' are embedded in a spherical 'pudding' of positively charged protons.

This delicious model was demolished in the early 20th century when Ernest Rutherford showed that the atom was like a miniature solar system with electrons orbiting around the central 'sun' or nucleus composed of protons and neutral particles called neutrons. After announcing his model, the world-famous professor of physics at Manchester University, with a broad grin and in a booming voice, said to his close colleagues of his critics: 'Some of them would give a thousand pounds to disprove it'. No one had the audacity—or a thousand pounds—to challenge the model that soon, with some changes, became the icon by which we still recognise the atom.

While students struggled to understand the three-particle structure of the atom, physicists came up with complex quantum models of the atom and discovered an entire 'zoo' of elementary particles (so many that it prompted Enrico Fermi to remark: 'If I could remember the names of all these particles, I would have become a botanist.').

The most famous of these particles are quarks, which were postulated in 1964 by the American physicist Murray Gell-Mann, who won the Nobel Prize in physics in 1969 for his work on them. Their name comes from a phrase—'Three quarks for Muster Mark'—in James Joyce's novel *Finnegans Wake*. Until recently, quarks were considered the basic building blocks of matter, but some physicists now believe that quarks themselves are made up of even smaller particles.

The physics of quarks and other elementary particles is very complex. In simple terms, we can say that each particle has three major characteristics: mass (some particles have zero mass); charge (every particle has a positive, negative or neutral charge); and spin (every particle spins like a top).

From matter to antimatter

If you are a *Star Trek* fan, you probably know that the starship *Enterprise is* powered by antimatter. Antimatter is not the stuff of science fiction; it does exist.

As early as 1898, Arthur Schuster, a British physicist, suggested the fascinating idea that an exotic type of matter could exist with properties that mirror those of ordinary matter. In a letter to *Nature*, he wondered: 'If there is negative electricity, why not negative gold, as yellow as our own?' He added that this speculation was just 'a dream.'

In 1928 the gifted British theoretical physicist Paul Dirac provided the mathematical basis for Schuster's dream. Dirac predicted that the electron, which is negatively charged, should have a positively charged counterpart: 'This would be a new kind of particle, unknown to experimental physics, having the same mass and opposite charge as the electron. We may call such a particle an anti-electron.'

The discovery in 1932 of the anti-electron (now known as the 'positron,' short for 'positively charged electron') in the cosmic radiation by the American physicist Carl Anderson vindicated Dirac's bold prediction. Twenty-three years later, scientists at the University of California at Berkeley created the anti-proton in a particle accelerator.

We now know that every fundamental particle has an anti-particle—a mirror twin with the same mass but opposite charge. The idea of anti-particles is now also applied to atoms—anti-atoms, which make up the antimatter.

When antimatter meets ordinary matter, they annihilate each other and disappear in an explosion in which mass is converted into energy as dictated by Einstein's famous equation $E = mc^2$, where E is energy, m is mass and c is the speed of light. The energy released in matter-antimatter annihilation is awesome: in a collision of protons and anti-protons, the energy per particle is close to 200 times that available in a hydrogen bomb.

If matter and antimatter annihilate each other, there is no likelihood of antimatter existing on Earth, or even in the solar system. The solar wind, the spray of charged particles emitted by the Sun in all directions, would annihilate antimatter. However, scientists speculate that antimatter could exist in other parts of the universe, but so far they have found no evidence. This has not stopped them from creating antimatter in the laboratory.

A team of scientists at CERN, the European particle physics lab in Geneva, did just that in early 1996. For about 15 hours they fired a jet of xenon atoms across an anti-proton beam. Collisions between anti-protons and xenon nuclei produced electrons and positrons. These positrons then combined with other anti-protons in the beam to make anti-hydrogen, the simplest anti-atom. Scientists could detect nine anti-hydrogen atoms. Hydrogen is the simplest (just one electron orbiting a single nuclear proton) and most abundant (it makes up about 75 per cent of the universe) of 118 chemical elements known to us. An anti-hydrogen atom would have a positron orbiting a single anti-proton.

'It's really the proof that there is an antiworld,' exulted the CERN team's leader, Walter Oerlert of the Institute of Nuclear Physics in Germany. Since 1996, CERN scientists have been regularly synthesising anti-hydrogen atoms and have so far collected several hundred thousand of them. This harvest would provide scientists with an insight into the properties of anti-hydrogen.

So now there is the experimental proof that antimatter does exist, what can it be used for? Because the annihilation of matter and antimatter creates enormous amounts of energy—hundreds of times as much as generated in a nuclear reaction—it is tempting to look at antimatter as a potential source of energy. This energy might one day provide the fuel for interstellar voyages, the same way matter-antimatter annihilation powers the fictional spaceship Enterprise. The amount of antimatter required for space flights is unbelievably small. A few hundred micrograms could fuel a spacecraft to Jupiter, and the round trip would take only a year.

If you find all this a bit too far-fetched, then what about the idea of an anti-universe—a universe parallel to ours. Enter it, and you will find your antimatter counterpart: anti-you. Don't shake hands—you'll annihilate each other.

Antimatter Tunguska

In 1940, when the idea of antimatter was nothing more than a mass of mathematical equations, the Russian scientist Vladimir Rojansky suggested the possibility of the existence in outer space of contraterrene meteorites (contraterrene, CT, or its phonetic transcription Seetee are obsolete terms for antimatter and were once very popular in science fiction; the ordinary matter was called terrene). Rojansky also said that such a meteorite 'would be entirely radiated away before reaching the sea-level'.

During the same year, *The New York Times* reported on 15 September: 'As the 22-foot cutter-type sailboat Rockit II was crossing Long Island Sound near Bridgeport, Conn., yesterday morning with four peaceful persons aboard, a shell screeched across her bow and exploded in the water 100 yards away'. A passenger recalled: 'It was a most disquieting experience ... The screech came first—an unholy noise. Then, a split second later, the explosion, about two points off the starboard bow. It blew up a great tower of water, twenty or thirty feet in the air. It was the strangest thing in the middle of the peaceful Sound. Why, there wasn't even a boat in sight! And not an airplane overhead!'

Authorities investigated the incident and found that no artillery shell could have exploded near the boat. As a meteorite on striking water would not explode, many astronomers of the time speculated that the explosion might have been due to the fall of a tiny contraterrene meteorite. In a comment on the *Rockit II* mystery in *Popular Astronomy*, Samuel Herrick Jr, an astronomer at the University of California, supported the contraterrene meteorite hypothesis and said that Dirac and other scientists 'are to be congratulated on one of the most ingenious (and entertaining) hypothesis of recent years'. He also warned his fellow astronomers that they 'will have to distinguish between the highly explosive fireballs or bolides from which no material reaches the ground, and which accordingly may be contraterrene, and those which are the source of terrestrial meteorites'.

This debate on contraterrene meteorites prompted Lincoln La Paz, a leading American meteorite expert who took a keen interest in Kulik's expeditions to Tunguska and co-translated many of his papers into English, to suggest in 1941 that the Tunguska meteorite was contraterrene because of the great amount of energy released, the absence of impact craters, and the absence of nickel-iron positively attributed to meteorites.

'If a contraterrene iron meteorite of a size comparable to those of the largest irons conjectured to have fallen should strike Earth,' he said, 'an extremely powerful explosion would result, since, in addition to the large store of heat energy resulting from the transformation of the kinetic energy of motion of the meteoritic mass, a vast amount of energy would be liberated by its annihilation.' He pointed out that no original meteorite material would remain at the site of the explosion.

Herrick and La Paz's explanations generated a somewhat angry response from Harvey N. Ninniger, a noted meteorite expert and the President of the American Society for Research on Meteorites. He said that both phenomena could be explained by demonstrated facts without assuming the existence of any such 'purely hypothetical material,' and 'we are surely courting a return to the days of "spirits and mystery" when we shrink from painstaking (or even back-breaking) investigations and seek refuge in untried hypotheses, especially when those hypotheses rest entirely on assumptions!' (In 1928, Ninniger had urged American scholarly associations to send an expedition to Siberia 'to secure what is yet available of this greatest message from the depths of space that has ever reached this planet'. No one showed any interest.)

When scientists as distinguished as Willard Libby, who had developed the carbon-14 dating technique, and his colleagues Clyde Cowan and C.R. Alturi suggested in 1965 that the Tunguska object had been composed of antimatter, they were probably not in danger of returning to the days of 'spirits and mystery'. Since Ninniger's warning in 1941 a lot had been discovered about antimatter. Although its existence had yet to be experimentally proved, antimatter was no longer considered a 'purely hypothetical material'.

In their detailed research paper in *Nature*, the three American scientists ruled out the possibility of nuclear fission or fusion reaction and argued in favour of the antimatter hypothesis. They said that neither fission nor fusion could explain the observed effects of the Tunguska explosion.

To start a fission chain reaction (in which a heavy atomic nucleus splits into lighter nuclei), a critical mass of a fissionable material such as uranium or plutonium is required. The multi-megaton blast at Tunguska would require a large initial mass—well above the critical mass—which seems unlikely. On the other hand, fusion (in which lighter atomic nuclei combine to form a heavier nucleus) requires a sufficient amount of tightly packed deuterium that must be heated to several million degrees Celsius. Such a high temperature could not be obtained just by entry into the atmosphere.

The antimatter hypothesis could explain the high nuclear energy yield of the Tunguska blast, but the researchers were quick to point out that 'several objections immediately arise' to this hypothesis. Two main objections were: (a) the lack of evidence for the existence of antimatter; and (b) the antimatter object would start disintegrating the moment it entered the atmosphere, and its largest yield of energy would be somewhere towards the middle of the path, rather than towards its end.

'A second look at the process tempers these conclusions, however,' they said. Of the three models for nuclear explosion, they decided in favour of the annihilation of an anti-rock in the atmosphere. Their calculations showed that if the Tunguska explosion had been due to an anti-rock, it should have behaved like a 35-megaton fission or fusion bomb. The explosion would also have generated trillions of radioactive carbon-14 atoms.

As the feasibility of the American trio's antimatter hypothesis depended upon the discovery of large amounts of radioactive carbon-14 in trees, they analysed carbon-14 content in sections of a 300-year-old fir tree that fell in 1951 in Tucson, Arizona, and an oak tree cut in 1964 near Los Angeles. They took nearly 90,000 counts of carbon-14 in tree rings from 1870 to 1930, which showed that the count peaked in 1909. However, the increase was much smaller than they had predicted. Their conclusion: although there are uncertainties, 'the data do yield a positive result'.

Recent measurements also show a rise in carbon-14, but not enough to support the idea of annihilation caused by a nuclear explosion, whether it be fission, fusion or antimatter.

A decade after the publication of the antimatter hypothesis, Hall Crannell of the Catholic University of America looked at other ways of measuring the antimatter content of the Tunguska object. He said that silicon, and to some extent aluminium, are abundant elements in rocks, and when the Tunguska anti-rock hit the ground, ordinary aluminium was converted into radioactive aluminium-26. If the aluminium-26 content of rocks or soil is measured as a function of the distance from the centre of the explosion, he suggested, the highest concentration of aluminium-26 should be found near the centre. No one has yet carried out such measurements.

The British astronomer David Hughes rejected the antimatter hypothesis on the ground that 'it is hard to understand how it penetrated to such a depth in the atmosphere and why the explosion maximised at the end of the trajectory and not midway along it'.

From antimatter to mirror matter

The idea of a 'mirror world' was first suggested in 1956 by Chinese American physicists Chen Ning Yang and Tsung Dao Lee. First, a tiny dose of particle physics, before you can enter their 'mirror world'.

The universe is held together by four types of fundamental forces—

gravity, electromagnetism, the strong force, and the weak force—which are transmitted or 'mediated' by the exchange of elementary particles. The gravitational force, or gravity, is the long-range force responsible for the attraction existing between all matter: it holds you to the ground and Earth in its orbit. Its range is infinite. The electromagnetic force is the attraction and repulsion between charged particles: it enables a light bulb to glow and a magnet to stick to your fridge. Its range is also infinite. The strong force is the 'glue' that holds together an atomic nucleus: it binds quarks to make protons and neutrons. The weak force is also a kind of nuclear force: it causes elementary particles to shoot out of the atomic nucleus during the radioactive decay of elements such as uranium. The range of the strong and the weak force is extremely short. The electromagnetic, the weak and the strong forces are very similar and are very well understood by physicists. However, gravity is still a mystery, and little is known of its relation to the other forces.

The existence of antimatter leads to the idea of symmetry; that is, every particle has a mirror-like twin. An anti-particle would look just like the ordinary particle, except that left would be switched with right. Physicists call it reversing the parity (parity is just a sexy word for left-right or mirror symmetry). Symmetry also applies to laws of physics such as the rules governing the interaction of elementary particles. All the original laws should continue to work in the same way: whatever could happen in the real world would also happen in the antimatter world.

But nature's symmetry is flawed. Certain interactions of elementary particles produce a particle spinning in the same direction. For example, when an atom emits a neutrino, it always spins in the same direction—left-handedly (if it were coming towards you, you would see it spinning clockwise). Reflected in a mirror, however, a neutrino would be right-handed (it would always spin anti-clockwise). In contrast, electrons can spin in both directions. As many elementary particles display a preference for left over right, the universe seems left-handed. Why? Physicists do not know.

In 1956 Yang and Lee suggested that the evidence for left-right symmetry was weak in interactions involving the weak force (which led Wolfgang Pauli, who had dreamed up neutrinos in 1930, to lament: 'I cannot believe God is a weak left-hander.'). This prediction was soon confirmed experimentally by other physicists. Mirror symmetry or parity was now dead. Asymmetry was the new king. The discovery of asymmetry won Yang and Lee the Nobel Prize in physics just a year later.

Like nifty accountants, Yang and Lee had to balance the books. They proposed a way to restore perfect left-right symmetry to nature: every right-handed particle might have a left-handed particle, and vice-versa. This means that in addition to the antimatter world, there might also exist a mirror world. In the mirror world, all neutrinos would be right-handed. Considered together, the real world and the mirror world would restore the symmetry that appears to be lacking in each.

Welcome to the mirror world—a world of mirror planets, mirror stars and even mirror life, all governed by mirror forces. This world is as fanciful as the one Alice entered *Through the Looking Glass*.

In this world, particles are right-hand or mirror images of ordinary particles. They also have the same mass as their ordinary counterparts. Thus, one force that acts on both ordinary matter and mirror matter is gravity. But there would not be any interaction between ordinary matter and mirror matter through nature's other three forces—the electromagnetic, the strong and the weak. We should be able to detect gravitational force when mirror matter comes near ordinary matter. The detection of this force would betray the presence of invisible mirror matter. The testability of this idea takes mirror matter out of the realm of science fiction into reality.

Because we are made of ordinary matter, we can neither see nor smell mirror matter (or our mirror matter twins, even if they were dressed in their brightest mirror matter clothes and soaked in mirror matter perfume). If you did encounter your mirror matter twin, you would pass right through him or her. You would also be invisible to your twin.

No mirror matter has yet been discovered or made in the laboratory, but neutrinos provide a misty glimpse of the mirror world. Neutrinos are the most pervasive elementary particles in the universe. There are about 50 billion neutrinos for every electron; they are everywhere but they cannot be seen and rarely interact with matter. Tens of thousands pass through our body every second. They have no charge and, although previously thought to have no mass at all, they are now believed to have a small amount of mass. There are three known types of neutrino—muon, tau and electron—and they are all created in the centre of Sun, in supernovas and in the cosmic rays hitting the upper atmosphere.

In his famous book *The Quark and the Jaguar*, Murray Gell-Mann writes that the neutrinos produced by the Sun 'reach the surface of Earth by raining down on us during the day, but at night they come up at us through Earth'. This aspect of neutrino behaviour inspired writer John Updike to write a poem entitled 'Cosmic Gall'. An excerpt:

> Earth is just a silly ball
> To them, through which they simply pass,
> Like dustmaids down a drafty hall
> Or photons through a sheet of glass.

Physicists have calculated the number of electron neutrinos that should reach Earth from the Sun. But they have detected fewer of these than predicted. Recent experiments have shown that neutrinos can change from one type to another. Some types of neutrinos are not spotted by neutrino detectors, which explains the discrepancy. The proponents of mirror matter solve the puzzle of missing solar neutrinos by suggesting the existence of a fourth type of neutrino—mirror neutrinos. These are so ghostly that they don't make their presence known to bewildered physicists.

There is more good news for those who believe in the existence of mirror matter. And it comes from a particle called orthopositronium. Positronium is like a hydrogen atom, but instead of an electron orbiting a proton, an electron orbits a positron, its antimatter counterpart. If the spin of the electron and the spin of the positron point in the same direction, the atom is known as orthopositronium. In 1986 Harvard physicist and Nobel

Laureate Sheldon Glashow suggested that orthopositronium could oscillate between mirror and ordinary orthopositronium—jumping back and forth through the mirror.

Orthopositronium is ephemeral; it lasts a mere 142 nanoseconds before its components annihilate each other in a burst of tiny energy in the form of three undetectable photons. However, in the 1990s, when physicists made a batch of orthopositronium, they found that its lifetime is shorter than 142 nanoseconds. In 2000, Robert Foot of the University of Melbourne and Sergei Gninenko of CERN suggested that mirror orthopositronium could explain the discrepancy. This could be due to orthopositronium changing fleetingly into its mirror matter form and then back again. The mirror orthopositronium would go undetected, and that could account for the shorter lifetime measurements.

Although mirror matter is expected to interact with ordinary matter only through gravity, recent experiments suggest a small electromagnetic attraction between mirror and ordinary particles. This coupling probably comes from the tiny electric charge that mirror electrons and protons are believed to have. This charge is about a millionth that of their ordinary counterparts. The tiny electromagnetic interaction between mirror particles and ordinary particles, if it exists, has interesting implications. It would make mirror stars visible if they had some embedded ordinary matter. This interaction would also be sufficient to heat up a body of mirror matter if it entered Earth's atmosphere. And that's where Tunguska enters the mirror world.

Mirror matter Tunguska

Robert Foot, who has been studying mirror matter since 1991, became interested in the Tunguska event when in 1999 he watched the television documentary *As It Happened: The Day Earth Was Hit*. He became convinced that the event was not fully understood by scientists and that they were ignoring the crucial evidence such as the funnel-shaped holes discovered by Kulik. He also thought it most unlikely that an ordinary matter asteroid or a comet could completely disappear in the air without leaving any traces, however minute.

In 2002, Foot proposed an interesting solution to the Tunguska puzzle. In his book, *Shadowlands: Quest for Mirror Matter in the Universe*, he suggested that the event was caused by a mirror asteroid. As it dived into the atmosphere the heat caused it to explode at high altitude. The explosion caused a shock wave that wreaked havoc on the Tunguska taiga but didn't leave a trace of an impact crater. He estimated that the mirror matter space-body was roughly 100m in size and weighed about one million tonnes. Such a heavy (ordinary or mirror) body would not lose much of its velocity in the atmosphere if it remained intact. However, if it were to break up for any reason, the energy of the body would be rapidly dumped into the atmosphere, leading to a huge explosion.

As for the nature of the space-body, Foot said that it was most likely made from mirror matter ices, such as mirror H_2O ice. An important difference between the mirror ices and the ordinary ices would be that the mirror ices would not be melted by light from Sun, and therefore would be likely to be relatively abundant in the inner solar system. On entering the atmosphere, a mirror H_2O ice body would vaporise during the flight, and any leftover fragments would eventually melt after striking the ground. 'This could explain why no substantial mirror matter fragments were found at Tunguska; most of the space-body had vaporized after it exploded in the atmosphere, any remaining fragments had melted by the time Kulik arrived there,' he said. 'Once in the liquid state, mirror matter should seep into the ground, probably making its extraction impossible.'

However, Foot has left some hope for Tunguska trophy hunters. The mirror body might have some embedded amount of ordinary matter, so a tiny amount of ordinary extraterrestrial material was possible. Also, any mirror matter fragments that survived and hit the ground could potentially cause small craters or holes. 'Perhaps the most interesting facet of this interpretation of the Tunguska event is that there should be large pieces of mirror matter still lodged in the ground at the Tunguska site,' he said. 'Nobody has looked.'

Perhaps the most spectacular way to test Foot's idea is to actually find mirror matter in the ground at Tunguska. 'Any mirror matter fragments would have melted when they hit the ground and reformed becoming mixed with ordinary matter at some distance underground,' he said. 'There may be some amount close to the surface which could potentially be extracted and purified'.

He suggested that the mirror matter might be separated from the ordinary matter in a laboratory centrifuge. But there is a catch: once the heavier mirror matter had separated from the ordinary matter, it would fly out of the centrifuge test tubes. As you can't see it, you can't catch it. The experiments, however, would prove the existence of mirror matter if the mass of the test tubes and their contents after the experiment were less than they were before. The missing mass would be the mass of the mirror matter you failed to catch.

'It would be a very exciting experiment and lots of fun too!' Foot believes. If you were ever interested in making a name for yourself as an experimental scientist, here's your opportunity. However, you must heed Foot's warning before you pick up your shovel and head to Tunguska: 'It is possible that mirror matter could be hazardous to health'. He takes no responsibility for any cases of mirror matter poisoning.

Alice (of the Wonderland fame) probably knew about mirror matter poisoning. Just before she stepped through the looking glass, she asked her cat, 'How would you like to live in a Looking-glass House, Kitty? I wonder if they'd give you milk in there? Perhaps Looking-glass milk isn't good to drink'. Now we know that this milk would be made of mirror molecules and perhaps it wouldn't be good to drink, for Kitty at least. Mirror-Kitty would love it, for sure.

There's another matter

And it's called quark matter. Quarks are incredibly tiny elementary particles. They come in six types or 'flavours': up, down, charm, strange, top and bottom. Protons and neutrons are made from up and down quarks. Other quarks are not found in ordinary matter. Scientists, however, believe that a strange quark matter—a form of matter made of up, down and strange quarks—was formed in the Big Bang that marked the beginning of the universe 13.8 billion years ago. This matter is so dense that a teaspoonful of it would weigh billions of tonnes.

In 2002, the orbiting Chandra X-Ray Observatory spotted a star which scientists believe is a quark star. Theorists have long suspected the existence of quark stars, collapsed stars that are denser than neutron stars but not dense enough to become black holes. The observed star has a radius of 5 to 6 km. This radius, scientists believe, is about half of what would be expected if the object were a neutron star, but about right if it were a quark star. If the strange quark matter does really exist, it could destroy ordinary matter by converting protons and neutrons to quarks. This process could spread like cosmic wildfire through space.

In the same year, a team of researchers from Southern Methodist University in Dallas said that not only does the strange quark matter exist, but it passed through Earth twice in 1993. The first event occurred on 22 November, when an object entered Earth off Antarctica and left the Indian Ocean south of Sri Lanka 26 seconds later. In the second event, on 24 November, an object entered south of Australia and exited near Antarctica 19 seconds later. These entry- and exit-points imply that both objects were travelling with a speed of 144,000 km/h. Both events were recorded by several monitoring stations, but no satisfactory explanation for the events has ever been suggested.

In 1984 Sheldon Glashow said that strange quark matter would pass through Earth with a dramatic effect: a 1-tonne object would release 50 kilotons of energy, which would be spread through Earth along the path of the object. The Southern Methodist researchers began looking for such events in 1993 and selected the two events mentioned above from more than a million records collected by the US Geological Survey between 1990 and 1993 which were not associated with traditional seismic disturbances. According to Eugene Herrin, one of the team members, normal earthquakes and explosions are point sources—the energy radiates from a single point.

When a clump of strange quark matter—known as a strangelet or strange-quark nugget—passes through Earth, you have a linear source with energy radiating from the whole of the line through Earth, and this would give a different pattern in the way seismic stations record the data. The 1993 events—caused by strangelets just one-tenth the breadth of a hair and weighing nearly a tonne—left a distinct linear quark matter pattern. 'We can't prove that this was strange quark matter, but that is the only explanation that has been offered so far,' Herrin said.

The impact of strangelets on an inhabited area would probably be less violent than that of a meteorite. 'It's very hard to determine what the effect would be,' said Herrin. 'There would probably be a tiny crater, but it would be virtually impossible to find anything.'

Did a quark matter rock strike Tunguska on 30 June 1908? There is a possibility, but no one has the definitive answer, yet.

Let's not be in the dark about the dark matter

In 1929 American astronomer Edwin Hubble suggested that galaxies are moving away from us and each other at an ever-increasing rate. More distant the galaxy, the faster it is moving from us. This means that the universe is expanding like a balloon.

The discovery that the universe is expanding was one of the great intellectual revolutions of the 20th century, remarks Stephen Hawking in his mega-selling book *A Brief History of Time* (1988).

Accepting that the universe is expanding, you might wonder whether it will go on expanding forever or whether the receding galaxies will someday stop and reverse their motion eventually. And then falling together in a great collapse—a Big Crunch, a kind of reverse Big Bang? All recent observations show that the universe will expand forever.

There exists more matter in the universe than we see as bright galaxies and bright stars. This so-called 'missing mass' comprises up to 95 per cent mass of the universe, and it is in the form of dark energy and dark matter. Roughly 70 per cent of the universe is dark energy; dark matter makes up about 25 per cent.

Scientists are still in the dark, pardon the pun, about dark energy and dark matter. But not C.D. Froggatt of Glasgow University and H.B. Nelson of The Niels Bohr Institute in Copenhagen. In 2014, they proposed that a dark matter ball caused the Tunguska event. The ball was less than a centimetre wide, but it weighed 100,000 tonnes. It hit the ground with a speed of about 160 km/s. The ball didn't stop near the ground like a meteorite, let alone in the air as it is speculated for a comet. Instead, it penetrated deep into Earth and only stopped after a few thousand kilometres.

As it moved, the ball heated up enormously a narrow tube. The heated matter in the tube came out like a volcanic explosion hundreds of kilometres under the surface. In short, an extraterrestrial object didn't cause the Tunguska explosion; it was a blast from below as described in various scenarios in the next chapter.

Chapter 8
A Blast from the Below

The line-up of suspects in the crime against 80 million trees of the Tunguska taiga so far includes a comet, an asteroid, a mini black hole and a rock of antimatter or mirror matter. Some contemporary scientists reject these extraterrestrial suspects because no definitive ET remnants have ever been found. They point the finger at something much closer to home. Their line-up of terrestrial suspects includes a massive gas explosion, a giant lightning ball and a so-called geometeor.

The escapade of a gas

Wolfgang Kundt, a German astrophysicist, dismisses comet and asteroid theories as pseudoscience and suggests an alternative 'volcanic blow-out' scenario for the fireball. As natural gas escaped from narrow underground volcanic vents, it became charged. The escaping gas, which contained mostly methane, raced upwards at high speed and started mixing with air. After a few seconds, this charged, volatile mixture sparked lightning that ignited it into a fireball. This massive fireball contained as much as 10 million tonnes of natural gas and caused the devastation we all know about.

The Tunguska site, according to Kundt, lies in the middle of the 250-million-year-old Kulikovskii volcanic crater, which has several faults or fractures running through layers of rocks. In the Tunguska blast, the gas escaped through a kimberlite, a conical volcanic pipe in the rock, which had been formed when hot magma from Earth's molten core pushed upwards under high pressure. Diamonds are formed in kimberlites more than 150 km beneath the surface and brought up in volcanic vents or pipes called diatremes. Kimberlites are named after Kimberley in South Africa, where legendary diamond reserves were found in the 1870s. Once thought to be common only in South Africa, kimberlites have now been found in other parts of the world.

The first expedition to the Tunguska site was carried out in 1910 by a wealthy Russian merchant and goldsmith named Suzdalev who, on return, urged the locals to keep silent about his expedition. Apparently, they obeyed. We do not know whether Suzdalev left Tunguska with his sledge filled with sacks of diamonds, but we do know from Kundt that the ejection of gas caused the formation of diamond-rich kimberlite diatreme pipes.

Kundt claims in the journal *Current Science* that dozens of funnel-shaped holes, including the famous Suslov crater discovered by Kulik, had been 'blown from below' during the morning of the explosion. He cites Evenk eyewitness accounts in support of his claim. He explains the presence of ice crystals in permanently frozen mud in the Suslov crater by saying that 'during its formation water should have intruded into its cavities'. Russian scientists have explained this phenomenon as the result of permafrost and have pointed out that similar holes are often found in other parts of Siberia.

Kulik had also found a rotten tree stump at the bottom of the Suslov crater. How could a tree stump make its way to the bottom of a crater blown from below? Kundt explains: 'There have been dozens of trees, standing on top of what is now the Suslov hole. Most of them were hurled to large distances, but one of them may have managed to fall back in, more or less upright. Alternatively, this stump could have slid back from the crater's rim along with transient mudflow.' Kundt believes that his theory has also answered the questions: 'Why did several expeditions find large numbers of detached tree stumps lying around in the Great Cauldron and its vicinity?'; and 'How did they get there?' He says: 'To me, they are a clear indication of ejections, from the holes at whose surfaces they had grown.'

The pattern of treefall in the blast area, which has been studied by many researchers, has many unique features. The almost radial pattern— 'wiggly rather than straight-line radial,' according to Kundt—has at least five centres and follows the valleys and hills. It shows islands of survived trees in the valleys. It also has 'telegraph poles' near the epicentre, which are reminiscent of the Hiroshima blast waves. Kundt claims that such fine structures of blasting and felling could not be explained by one big explosion near the ground. He considers the impact models inconsistent, as 'all of [them] produce parallel treefall patterns if properly evaluated'. The actual pattern 'requires several successive localised explosions near the ground,' he says. A detailed study of the treefall pattern made by the 1991 Italian expedition has also suggested more than one centre of the explosion.

Kundt explains 'bright nights' observed in parts of Europe and Asia after the Tunguska explosion by saying that 'such nights are unique in the recorded history except for the 1883 Krakatoa volcanic eruption'. He says that the bright nights of Krakatoa and Tunguska were due to natural gas—mainly methane—which could rise supersonically to a height of 200 km. There it was reheated by solar radiation and by slow-burning with the surrounding atomic oxygen and rose again to an altitude of about 600 km. Water vapour formed during slow-burning froze out as snowflakes, which scattered Sunlight and gave rise to bright nights.

Kundt's other reasons against the extraterrestrial origin of the Tunguska blast include:

- An asteroid would have left a trace, whereas a comet would have disintegrated too high, and also would have been discovered before the impact.
- The absence of any remnants of the interstellar body. An iron comet would have left a large, lasting crater. A stony asteroid would have left either big fragments or at least 4-mm-thick dust layer.
- Several witnesses reported the sound of gunfire before they saw a 'pillar of fire' in the sky. This order of events is expected from a volcanic blow-out, not from an extraterrestrial impact.
- The heat felt by many witnesses at Vanavara, about 70 km from the explosion site, cannot be explained by a meteorite trail. A meteorite trail cannot produce such intense heat because it is narrow and short-lived; it would have to pass quite near to Vanavara with a rather large speed. 'What counts is the spherical

angle of the hot source, seen by your face: you feel the heat of a near bonfire—covering a large spherical angle—but you cannot possibly feel the heat of a (short and narrow!) meteoritic trail,' he says.

For his last reason—number 26 on his list—Kundt relies on statistical odds. It is well known to geologists that only a small number of terrestrial craters were gouged in by rocks from space; most of them were formed by volcanoes. In all the faces of volcanism—ranging from hardly noticeable outgassing, through lava flows, to mud volcanoes, real volcanoes and explosive, supersonic ejections—rising natural gas is the primary piston.

There are other reported cases of natural gas explosions, but none as dramatic as the Tunguska blast. In his *Current Science* paper, Kundt quotes a 1988 incident reported to him by the American geologist Thomas Gold: 'A United Airlines plane on the way from Tokyo to Honolulu in calm air experienced a sharp upward bump followed, in a fraction of a second, by a mightier downward movement with a recorded speed implying a downward acceleration of 4 g'.

Gold explained the upward bump as the crossing of a methane cloud rising at high speed. The plane's engines then ignited the methane-air mixture above the aircraft. This explosion forced the plane downwards and injured many people seriously. The plane had to return to Tokyo to attend to the wounded.

Kundt considers Tunguska 'a lovely detective story which requires a broadly educated physicist's mind for its resolution'. He says that he owes a lot of thanks to Andrei Ol'khovatov (see 'Ghostly geometeors' below), whose sober analysis converted him from the mainstream opinion to a physically consistent one.

After Kundt's presentation at a conference on environmental catastrophes in London in 2002, Jesus Martinez-Frias of the National Aerospace Institute in Madrid (obviously with a 'broadly educated physicist's mind') said that Kundt's geophysical hypothesis was 'a fresh idea'. 'It could be the answer,' he suggested.

Like Kundt, Vladimir Epifanov, a geologist from the Siberian Research Institute of Geology, Geophysics and Minerals, also believes that the epicentre of the Tunguska explosion is indeed located just above a major oil and gas field. But he suggests a different mechanism for the explosion: a powerful fluid jet that had suddenly shot up under high pressure from the depths of Earth.

According to Epifanov, gases from the oil deposits and methane from the coal beds, which had accumulated under a thick cover of basalt, suddenly broke through one day. A moderate earthquake could have initiated the process. The fluid jet was accompanied by dust that created a layer of aerosols in the upper atmosphere. If this layer became charged with electricity, it could have produced the spark that set off the explosion. The fireball then rushed towards the ground, flattening trees in a circular pattern for many kilometres. However, the cooling as the escaping gas rapidly lost pressure could have formed an ice dome around the place where the gas discharged, protecting the trees in the centre of the blast. This would not be the case with an impact from space.

Critics of the 'blast from below' scenario ask: what about the fireball that was seen by many witnesses racing across the Siberian sky from the south-southeast to the north-northwest just before the explosion? A volcanic gas blast fails to explain a racing fireball in the sky. Have these eyewitness accounts flown out of the window—just like a ball of lightning?

A giant lightning ball

A while ago, a reader wrote to the 'Science Times' section of *The New York Times* about a lightning ball that was seen by her family to 'enter the glass front door, go right past us (or possibly even through us) in the living room and leave by the back window, where it hit a tree, causing some damage'. The Science Times commentator joked, 'Next time, take a picture,' because it is one of nature's rarer phenomena and few photographs of it exist. It is also the least understood. Ball lightning has attracted the attention of scientists for two centuries, but it remains an enigma—dismissed by many as a myth or an optical illusion.

Over the years, scientists have collected thousands of accounts of sightings of ball lightning. In 2002 the Royal Society's journal *Philosophical Transactions* presented a selection of recently reported sightings. One account describes a lightning ball as it entered through an open window in the pantry of a house in Johannesburg: 'It entered the kitchen around the corner then sped out of the kitchen again around another corner and into the passage and the hall where it hit the tin bucket with a clang! Certainly, when we ran to check, the bucket was too hot to pick up, and its paint had blistered!'

In another account, a white-grey lightning ball 70 to 80 cm in diameter and with the glow of an incandescent lamp of 200 watts bounced on the head of a Russian teacher who was with her friends: 'It appeared as if from nowhere. We got frightened, squatted, and connected our heads, creating a circle. The ball suddenly began to move over us in a circle, and it also moved up and down. It was at a height of 0.5 m above the ground. Then it "chose" my head and began to jump on it, up and down, like a ball. It made more than 20 jumps. It was as soft as a bubble.'

The journal also listed an extraordinarily large—about 100 m in diameter—lightning ball that was caught on colour film by a park ranger in Queensland, Australia. It was anchored to the ground and lasted surprisingly long, about five minutes.

One of the rare accounts of ball lightning witnessed by a respected scientist comes from the British radio astronomer R.C. Jennison, who encountered a lightning ball on a late-night flight in stormy weather in 1963. He described it in a letter to the journal *Nature* in 1969 as a blue-white glowing sphere a little more than 20 cm in diameter which emerged from the pilot's cabin and passed down the aisle of the aircraft approximately 50 cm from him, maintaining the same height and course for the whole distance over which it could be observed. It then passed through the metal wall of the aircraft.

From such accounts, scientists have painted a picture of this bizarre phenomenon, which is always observed during stormy weather. A lightning ball is usually seen as a free-floating, luminous sphere that shines for a few seconds to a few minutes before it either explodes with a sharp bang or flicks out in silence. It can be almost any colour, sometimes even a combination, but green and violet are rare. Its size varies from a small ball to a giant globe several metres in diameter. It may suddenly appear in the air, or even from holes in the ground, chimneys, sewers and ditches. It usually moves horizontally in the air (at speeds between 3.5 and 350 km/h) about a metre above the ground. Still, it can climb utility poles and then dart along power or telephone lines. It can even dive down chimneys and squeeze through spaces much smaller than its size, but it never changes its size. It seems cool to the touch, but it may destroy electrical equipment, melt glass, ignite fires and scorch woods or singe people and animals. Sometimes a hissing or crackling noise can be heard. It may leave behind a sharp and repugnant smell, resembling ozone.

Ball lightning has been the subject of serious scientific research since the early 19th century, but no consensus theory has yet emerged. One of the popular theories is the plasma theory which says that a lightning ball is a sphere of plasma or a hot gas of electrons and positively charged ions. Another theory that is gaining favour comes from New Zealand scientists John Abrahamson and James Dinniss. When lightning strikes soil, it turns silica in the soil to pure silicon vapour. As the hot vapour cools, the silicon condenses into a floating ball of silicon aerosol held together by electrical charges. The chemical energy stored in silicon is slowly released as heat and light. Because the ball would become visible only over the latter part of its lifetime, it would appear to materialise out of thin air after a lightning strike. So simple, yet so amazing! Unfortunately, most of us will never see it (about one per cent of the population will see ball lightning in a lifetime).

Like Tunguska, ball lightning is a favourite of both scientists and charlatans. Is there a link between Tunguska and ball lightning? Key in 'ball lightning + Tunguska' in Google's search engine and you will be instantly presented with a list of thousands of web pages. A close analysis shows that most of these are alternative science pages discussing Tunguska and ball lightning in a similar vein as UFOs and alien abductions. The scientific link between Tunguska and ball lightning seems tenuous.

In his book *Cauldron of Hell: Tunguska* (1977) the American science writer Jack Stoneley poses the question: could some particularly massive form of ball lightning conceivably be associated with the Tunguska event? To answer it, he quotes the British scientist Anthony Lawton, who was also the scientific editor of Stoneley's book, as saying that to cause such devastation would require a lightning ball nearly one kilometre in diameter. Stoneley claims that from eyewitness reports early Tunguska researchers reckoned the fireball to be about one kilometre across. 'This is so close that we cannot dismiss the possibility that the Siberian monster might have been a giant lightning ball,' he writes.

Can a lightning ball grow to such an enormous size? 'Lawton maintains it is possible, provided the ball lightning is composed in a particular way,' Stoneley writes. 'He suggests that if the fireball is made up from dust particles bound tightly together by an electric charge, it could achieve these huge dimensions.' Lawton's prediction now sounds believable in view of the new aerosol theory of ball lightning.

David Turner, a British physical chemist now retired and living in the United States, looked at Tunguska and ball lightning in a different light. In an exhaustive analysis in *Physics Reports* he says that various studies of Halley's comet in 1986 found the temperatures in the plasma region of the comet to be much higher than expected. This looks like an exact analogy of the high temperatures implied for lightning balls, he says. He then went on to say that these observations could have relevance to 'one of the most spectacular and perplexing events' in the 20th century.

Turner lists five factors that do not support the asteroid theory:

- The very high percentage (more than 10) of the energy released as electromagnetic radiation.
- The occurrence, six minutes after the explosion, of a local magnetic storm that lasted more than four hours.
- Optical anomalies (bright nights and so on) seen in some parts of Europe and Asia, which began a week before the event but peaked on the morning of the explosion.

- The failure to find virtually any fine material on the ground that can be reliably associated with the impacting body.
- An apparent change in direction of the falling body (both in the horizontal and the vertical plane) which occurred shortly before the explosion.

For Turner the most important outstanding question is the determination of the maximum temperature sustainable within ball lightning. Present estimates vary from 400 to 15,000°C. He believes that this is closely related to the question of whether the Tunguska event could have resulted from a natural hydrogen bomb explosion. He is implying that the plasma in the comet, which was like a lightning ball, was hot enough to start a hydrogen bomb reaction. 'It may be premature to discount the comet hypothesis at this stage,' he advises.

G.G. Kochemasov of the Russian Academy of Sciences believes that partisans of the comet and asteroid theories fail to consider two important points. First, the anomalous atmospheric conditions long before the event. Second, the non-linear motion of the object. These two reasons are similar to those listed by Turner. Kochemasov says that a giant lightning ball can explain these two anomalies. Earth's restless ionosphere causes various electrically charged events such as aurora borealis. As flights of ball lightning have been noticed along geomorphological boundaries, it is possible that a giant lightning ball formed in the Tunguska region, which is in an area of volcanic and tectonic activities. Thus, it was a home-grown product without any ET connections.

Kochemasov estimates the diameter of the lightning ball to have been about 200 m. Such giant balls have never been observed but, he believes, we have to think in terms of thousands of years, or geological timescale. He cites two historical occurrences of giant ball lightning: the archaeological evidence suggesting that the ancient Indus Valley city of Harappa (now in Pakistan) was ruined after an enormous fire; and mention in the ancient Indian epic *Mahabharata* of an 'explosion' that caused 'dazzling light, a fire without smoke'.

Kiril Chukanov, an independent researcher on ball lightning in Salt Lake City, Utah, also believes that the Tunguska fireball was an enormous lightning ball about 500m across. On his 'Chukanov Quantum Energy' website and in his self-published book, *Final Quantum Revelation*, he lists nine reasons in support of his hypothesis. They include:

- The 11-year sunspot cycle peaked at the end of June 1908. This sunspot activity manifests itself in Earth's atmosphere by intensified geomagnetic activity and the presence of abnormal optical events such as bright night skies.
- Ball lightning explodes because of the leakage of electric charges from the sphere and the resulting disintegration of its structureless nuclear component.
- Ball lightning typically disintegrates into smaller spheres, which further disintegrate into still smaller spheres, until finally they explode. Eyewitness accounts of many explosions and simultaneous fire breakouts in widely scattered areas of the forest support a disintegrating ball-lightning scenario.
- Other theories fail to explain the enormous amount of energy accumulated for a short period on the surface of the Tunguska object.

Chukanov also believes that ball lightning can be used to create 'free energy'. The American Physical Society's Bob Park has labelled Chukanov's ideas as 'voodoo science'.

Ghostly geometeors

Andrei Ol'khovatov's geometeor hypothesis is definitely not voodoo science; however, some of his ideas seem ahead of their time. Ol'khovatov, formerly of the Soviet Radio Instrument Industry Research Institute and now an independent researcher based in Moscow, is a popular figure in the large Tunguska cyberspace community, as well as in the small but real community of Tunguska scientific researchers. His website and discussion forum keep the Tunguska debate alive and up to date.

Ol'khovatov became interested in the Tunguska event when in the late 1980s he read about earthquake lights, a glow that sometimes occurs before a large earthquake. He immediately associated these lights with the Tunguska eyewitness accounts. The similarity between accounts of earthquake lights and those of Tunguska led him to believe that there might be some link between the two. He first published his hypothesis in the journal of the Izvestia Academy of Science of the USSR in 1991, but an updated English version appears on his website (olkhov.narod.ru) and in the proceedings of many scientific conferences attended by him. Like all those who reject a cosmic impact, he wants to know where the remnants are. 'Nowhere,' he says, 'nothing after decades of detailed research'.

He believes that the explosion was caused by a strong coupling between some unknown subterranean and atmospheric processes. This coupling formed meteor-like luminous objects but of terrestrial origin. For want of a better word he calls these objects 'geophysical meteors' or 'geometeors'. A geometeor resembles high-speed ball lightning. 'Similar events occur in association with earthquakes (earthquake lights) and in association with a thunderstorm (ball lightning),' he says.

According to Ol'khovatov, the Tunguska region is right in the middle of an ancient volcanic crater. There are many other prominent geological faults, circular structures and other geological formations in the region. Several tectonic faults intersect near the Tunguska explosion centre. There is evidence of increased seismic activity in the region before the explosion. Simultaneously, there was also an increase in unusual meteorological phenomena: increased sunspot activity; sharp increase in thunderstorms; the change in Tunguska region meteorological stations' forecast for 30 June 1908 from 'good weather' to 'bad weather' because of the possibility of a cyclone; and a strong upsurge in atmospheric pressure soon before the event.

This rare combination of large-scale geophysical and meteorological disturbances manifested itself as follows. First, there was luminous activity in the atmosphere in southern Siberia, which was like falling meteors. At about the same time, a swarm of shallow earthquakes started, which were accompanied by brontides (thunder-like sounds of short duration believed to be of seismic origin). Then, at the vent of the crater, there was a large geometeor explosion.

Ol'khovatov believes science is not yet ready to explain the exact mechanism of geophysical and meteorological interactions. However, in his paper, he describes in detail how a geometeor can explain various eyewitness accounts and the natural phenomena associated with the event. Ol'khovatov quotes Russian researchers who in 1988 analysed eyewitness accounts of the Tunguska object and found the following descriptions for the shape of the object:

The shape of the Tunguska object was like a …	Percentage of eyewitness accounts …
Ball	18.8
Cylinder	16.3
Cone	2.1
Star	3.4
Tail	14.0
Snake	2.3
Lightning	2.1
Strip of light	2.5
Pillar of fire	4.9
Flame	10.3
Sparks	11.2
Other shapes	12.1

He points out that these descriptions hardly conform to a meteorite fall. He also explains the 'three trajectories of the Tunguska meteorite fall' drawn up by researchers from eyewitness accounts by saying that all three trajectories are above the main tectonic faults and they intersect at a point near Vanavara. Does it mean that witnesses saw more than one luminous phenomenon or geometeor? Ol'khovatov says: 'I'm inclined to think that there were several low-altitude fireballs, and that's why there are no reports of two or more fireballs seen simultaneously. The low altitude of fireballs explains why nobody in Vanavara saw a fireball or its trail. Besides the fireballs, evidently there were other typical earthquake lights.'

The three trajectories, according to Ol'khovatov, vary from south-southeast to east-northeast, that is, by up to 90 degrees. These trajectories were drawn from the accounts of eyewitnesses who were 500 km away from the epicentre. 'If it were a meteorite most of the witnesses from the west of the trajectory would say that it flew from the east, while most of the witnesses from the east of the trajectory would say that it flew from the west,' he says. 'There would be some witnesses who would say it flew overhead. So, we should have a well-defined trajectory. But there was no such situation in Tunguska. The advocates of the asteroid/meteorite impact choose just a small part of the eyewitness accounts and then announce other eyewitness accounts as "unreliable". As there are several trajectories, so each trajectory has its "reliable witnesses" (a minority), and its "unreliable witnesses" (a majority). A witness for one trajectory could be "reliable" and "unreliable" for all other trajectories.'

At a qualitative level, says Ol'khovatov, his geometeor hypothesis can explain all known facts about the phenomenon. For example, a few years ago several articles appeared in scientific journals stating that about a day or less before an earthquake, the cloudiness level sharply decreases above tectonic faults in the area that will become the epicentre of Earthquake. 'So I studied data from nine meteorological stations within 1,000 km of the Tunguska epicentre (the closest is 500 km away),' Ol'khovatov says. 'I averaged the data from all stations. And indeed, the average daily cloudiness level shows an extremely deep drop on 29 June 1908!'

Ol'khovatov estimates the diameter of the largest ball to be about one kilometre. 'But it is just a guess,' he adds. He also stresses that the luminosity of the fireball reported was rather weak. At the same time, according to the meteorite theories it must be as bright as the Sun, and much brighter near the epicentre—with no persistent trail. He has not estimated the energy of the fireballs: 'The question is still open: Is energy deposited by a lightning ball inside it, or does it also include energy around it? In ball lightning, what we see is a tip of the iceberg. I believe science is not yet ready to give a final answer to this question.'

We shall wait. But for many other explanations for the Tunguska mystery, we do not have to wait. Science is ready right now to judge them. Whenever the word 'mysterious' is attached to a natural scientific phenomenon that lacks a beyond-a-reasonable-doubt explanation, it becomes a fertile ground for the imagination of those who love outlandish ideas. There is no shortage of eccentric theories for the explanation of the Tunguska event. Let's open the X-files.

Chapter 9

Opening the X-files

FBI agents Fox Mulder and Dana Scully investigated their Tunguska mystery in *The X-Files: Tunguska*, in which Mulder encounters a deadly form of alien life form, 'black oil', living in rocks mined from a giant impact crater in Tunguska. But in the following files, dear reader, you're on your own.

Spaceship Tunguska

In August 1945, Little Boy and Fat Man changed our world forever. Curious names for two atomic bombs that unleashed untold devastation upon humanity. On 6 August, Little Boy almost wiped the city of Hiroshima from the map of Japan. Three days later, Fat Man exploded into history in the Nagasaki sky.

'It was hard to believe what we saw,' said Colonel Paul Tibbets, the pilot of the B-29 plane that dropped the atomic bomb on Hiroshima, describing at a press conference what he saw seconds after the bomb had been released. 'Below us, rising rapidly, was a tremendous black cloud ... What had been Hiroshima was going up in a mountain of smoke. First, I could see a mushroom of boiling dust—apparently with some debris in it—up to 20,000 feet. The boiling continued three or four minutes as I watched. Then a white cloud plumed upward from the center to some 40,000 feet. An angry dust cloud spread all around the city. There were fires on the fringes of the city, apparently burning as buildings crumbled and the gas mains broke.'

On the ground, Kiyosi Tenimoto, a pastor of the Hiroshima Methodist Church, who was about 4 km from the centre of the explosion, saw a blinding flash of light, like 'a sheet of sun,' that cut across the sky. Moments later, the flash of light had turned into a giant mushroom cloud, now known to everyone as the characteristic signature of an atomic explosion. John Hersey, one of the first Western journalists to record the bomb's immediate aftermath, reported in *The New Yorker* magazine of 31 August 1946 that the survivors described the explosion as 'a noiseless flash of light'. He noted that almost no one in Hiroshima recalled hearing any noise of the bomb, but all saw the vast, blinding glare and felt the wave of heat, which was followed closely by the roar of the explosion and its shock.

Hersey's excellent article, 'Hiroshima'—published simultaneously as a Penguin book which remains in print—had a profound effect on a world which knew hardly anything about the horrors of the atomic bomb. A paragraph about the experience of Father Wilhelm Kleinsorge, a German priest of the Society of Jesus, alludes to the Tunguska fireball: 'After the terrible flash—which, Father Kleinsorge later realized, reminded him of something he had read as a boy about a large meteor colliding with Earth—he had time (since he was 1,400 yards from the center) for one thought: A bomb has fallen directly on us. Then, for a few seconds or minutes, he went out of his mind.'

As for the Hiroshima bomb's incredible destructive power, the numbers speak for themselves. The air temperature at the point of explosion of the 15-kiloton bomb, 580 m above the ground, exceeded one million degrees Celsius. The temperature on the ground at the centre of the blast rose to 6,000°C. The brilliant orange mushroom cloud climbed to 10 km. As the cloud spread, it started fires that damaged more than 70,000 houses and killed 140,000 people. But the death toll reached 200,000 due to radiation sickness. In short, two-thirds of an 18-square-kilometre city of 340,000 people were almost obliterated by one atomic bomb in a few minutes.

The world now knew of the immense destructive power of 'the fireball, the mushroom cloud and the intense heat' of an atomic bomb blast. It did not take some Soviet scientists and science fiction writers long to connect the images of Hiroshima and Nagasaki with the images of Tunguska—the fireball, the heat, the thundering noise, the enormous cloud of dust and the devastated forest.

One of them was Alexander Kazantsev, an engineer who had

graduated from Siberia's Tomsk Technological Institute in 1930. He was also a well-known science fiction writer, who in 1946 published a story, 'The blast', in *Vokrug Sveta*, a popular Russian magazine of science and adventure, in which he presented the bizarre idea that the Tunguska explosion was caused by a 'cosmic visitor'—an extraterrestrial spaceship, cylindrical and propelled by nuclear fuel. Because of a malfunction, the spaceship plunged out of control through Earth's atmosphere, and within a fraction of a second, it and its occupants were vaporised in a blinding flash of light. The ETs had come to collect water from Lake Baikal, 800 km from the Tunguska explosion site. This lake—the world's deepest (1,637m) and the seventh largest (34,000 square km)—contains the largest volume of surface freshwater. Apparently, the ETs were from a parched planet and very, very thirsty.

Kazantsev's ET hypothesis was well received by other science and science fiction writers. Over the years, as Tunguska expeditions published their new findings, Kazantsev returned to his story again and again and embellished it into a working theory to explain the Tunguska object. When Florenskiy's 1958 expedition announced the discovery of magnetite globules containing nickel, cobalt, copper, germanium and other elements in the samples collected from the region of the fall, Kazantsev was quick to explain the presence of these elements.

In his 1958 article 'Visitor from the Cosmos' (which became the centrepiece of his 1963 book of the same name), he said that the nickel and cobalt came from the outer shell of the spaceship, while the copper and germanium were from semiconductors and other electrical instruments on board. These and other elements were vaporised when, at the moment of the explosion, temperatures rose tens of millions of degrees. 'In part, these elements fell to the ground as precipitation, with radioactive effects,' he maintained.

The *Soviet Astronomical Journal* panned Kazantsev's book as 'a consistent and conscious deception of the reader, in pursuit of one definite goal: to show that he alone, A.N. Kazantsev, has discovered the true nature of the complex phenomena contrary to all the "conjunctures" of the representatives of official science'. But there were admirers as well. One of them was an aircraft designer, A.Yu. Manotskov, who 'proved' that the Tunguska object was under 'intelligent control'.

On average, a meteorite or comet would enter Earth's atmosphere at a speed of 36,000 to 216,000 km/h, whereas the Tunguska object 'braked' its speed to 2,400 km/h, the speed of a jet aircraft. For a meteorite or comet to plunge at this low speed, it would have needed a mass of 1,000 million tonnes and a diameter of one kilometre. Yet this behemoth had made no crater and left no fragments. Therefore, the Tunguska object was a small spaceship that was attempting to land. Kazantsev gleefully agreed: 'Such a tremendous meteorite would have certainly covered the whole sky.' Boris Laipunov, a well-known rocket and space travel expert, also supported Manotskov's reasoning.

Feliks Zigel of the Moscow Aviation Institute added more meat to Manotskov's assertion that the Tunguska object was under 'intelligent control'. Some eyewitness accounts, taken down long after the event, suggest that the Tunguska body had twice changed course in flight. This deliberate 'manoeuvre' to change course before descent was indeed proof that the Tunguska object was a spaceship flying from another planet, the good professor declared. He presented another 'proof': the object flew in an 'enormous loop', first northward then westward, before crashing; behaviour that appears to exclude a natural phenomenon. He said that the spaceship had followed precisely the re-entry angle of 6.2 degrees to the horizon, which was within the re-entry corridor (between 5.5 degrees and 7.5 degrees) adopted by astronauts entering Earth's atmosphere. If the angle is too steep, the spacecraft burns up; too shallow, and it bounces off the atmosphere like a stone skipping off the water.

In an article in the magazine *Znanyie-Sila* in June 1959, Zigel, who is still remembered as 'the father of Soviet UFOlogy,' heaped praise on Kazantsev's hypothesis: 'At the present time, like it or not, A.N. Kazantsev's hypothesis is the only realistic one insofar as it explains the absence of a meteorite crater and the explosion of a cosmic body in the air … It is generally known at times—nay, often—new ideas that proved to be most valuable to science were first expressed not by scientists, but by writers of scientific fantasy'. In an interview with the Soviet news agency TASS, he added: 'The more we know of the Tunguska catastrophe, the more confirmation we find of the fact that the UFO which exploded over the forest in 1908 was an extraterrestrial probe.'

The newspaper *Pravda* of the time, however, considered UFOs 'flirtations with superstitions and religious impulses manipulated indirectly by the Pentagon'. The reference to the Pentagon probably came from the Roswell incident, one of the most famous UFO 'sightings' in American history.

There's a flying saucer in my backyard

On 8 July 1947, the *Roswell Daily Record* broke the news of a cosmic encounter in New Mexico. The story, headlined, 'RAAF captures flying saucer on ranch in Roswell region,' was based on a press release issued by Roswell Army Airfield (RAAF). When sheep rancher Mac Brazel was making rounds at a ranch 137 km west of Roswell, he found some wreckage consisting mostly of rubber strips, wood sticks, tinfoil, plastic, tape with some strange markings that resembled 'hieroglyphics', and very tough paper. The unusual nature of the debris struck Brazel. After a few days, he drove into Roswell, where he reported the incident to the Sheriff, who reported it to Major Jesse Marcel, Intelligence Officer at RAAF.

The Army closed off the debris site while the wreckage was being cleared. The officers thought that they had found a flying saucer. They shipped the debris to Air Force General Roger Ramey for examination. But in the meantime, Colonel William Blanchard, the Commander at Roswell, issued a press release stating that the wreckage of a flying saucer had been recovered. The news caused a sensation around the world, but it was short-lived. Within hours General Ramey called in the local press and announced that RAAF had mistakenly identified remains of a weather balloon as the wreckage of a flying saucer. The next day, the *Roswell Daily Record*'s banner headline proclaimed: GEN. RAMEY EMPTIES ROSWELL SAUCER.

Another story that was not published in the paper, but that some town folks knew about, came from witnesses who had seen the wreckage. They claimed that they had seen alien 'bodies' nearby, describing them as a little more than a metre tall, with bluish skin colouration and no ears, hair, eyebrows or eyelashes. The Air Force explained these 'aliens' as dummies dropped from high-altitude balloons to study the results of the impact. And that was the end of the excitement.

No one ever talked about this episode, at least until the publication in 1980 of a book, *The Roswell Incident*, which came to the dramatic conclusion that there had been a cover-up of cosmic proportions. In 1988 another book, *UFO Crash at Roswell*, claimed that the US government had found and removed the remnants of the UFO crew—several little alien bodies. These two books were the genesis of a UFO myth and a conspiracy theory—that the government had conspired to cover up the fact that an alien ship had landed at Roswell. The truth is far less exotic: what happened was that people who saw the dummies mistook them for aliens.

The term 'UFOs' (unidentified flying objects) was suggested in the mid-1950s by the US Air Force. The term 'flying saucer' was not considered accurate, since many sightings had very natural explanations, while others did not. UFO fans tend to forget that the 'U' in UFO simply stands for 'unidentified,' and does not suggest 'extraterrestrial'.

The question of the existence of extraterrestrial life (even the simplest form, such as microscopic organisms) and alien intelligent life (advanced technical civilisations capable of communicating with us) is not related to UFOs. Since the dawn of history, humans have been pondering the question: are we alone?

If we are not alone, then how many intelligent civilisations might exist among the stars? In 1961 the American radio astronomer Frank Drake came up with an ingenious approach—now known as the Drake Equation—to answer this question. At that time, he was working at the National Radio Astronomy Observatory in West Virginia. In the early 1960s, many scientists ridiculed the idea of extraterrestrial intelligent life, but to Drake, the idea of other intelligent civilisations beyond Earth was a distinct possibility. He even placed a sign on his office door: 'Is there intelligent life on Earth?' He was serious about the search for extraterrestrial intelligent life, and those who questioned his belief in it were not, in his view, 'intelligent life' on Earth.

In its simplest form, the Drake Equation works as follows. To find out the number (N) of advanced technical civilisations in the Milky Way, we need to know:

- How many stars are born each year in our galaxy (R)
- How many of these stars have planets (p)
- How many of these planets are suitable for life (e)
- On how many planets life actually appears (l)
- On how many planets life evolves to an intelligent form (i)
- On how many planets the intelligent life can communicate to other worlds (c)
- The average life of these advanced civilisations (L)

If we multiply these seven factors, we get the equation:

$N = R.p.e.l.i.c.L$

If we know the values of these factors, we can calculate N. As astronomers do not agree on the exact values, estimates of N vary from one (we are home alone) to about 250 million (yes, a flying saucer could land in your backyard). These estimates are for our galaxy alone, and there are 200 billion galaxies (and still counting) in the presently observable universe. Mind-boggling, indeed!

If there are millions of ET civilisations in the universe, then it is possible that some of them might have visited Earth in the past. Drake, who is still involved in the search for extraterrestrial intelligence at the SETI Institute, says: 'As strongly as I believe that intelligent life exists elsewhere in the universe, I maintain that UFOs are not extraterrestrial visitors. They are the products of intelligent life on this planet.'

Zapped by a laser

Intelligent life on this planet in the form of two Russian science writers has suggested that Tunguska was mistakenly zapped by a laser sent by ETs from a giant planet orbiting the star 61 Cygni, about eleven light-years away from us. In a lengthy article published in the magazine *Zvezda* in 1964, Genrikh Altov and Valentina Zhuravleva said that the violent volcanic eruption of Krakatoa in August 1883 generated strong radio waves. This signal was received eleven years later by Cygnian scientists. But they misread the signal as greetings from a distant civilisation.

Following the age-old Cygnian custom, the courteous scientists decided to send a return message. As their laser technology was more advanced than the radio technology, they directed a laser beam at our planet. Unfortunately, the well-meaning scientists made another mistake. This time they misjudged Earth's distance and fired a powerful laser beam that zapped the Tunguska taiga. This 'extra strong' Cygnian message was all Greek to the local Evenk people; they did not have the required technology to read their greeting card from the stars.

How could a volcano generate a radio signal? Altov and Zhuravleva simply said that the volcanic ash flung high into the atmosphere, disturbed the ionosphere, which could have produced a radio signal. The signal was so strong that it reached far out into space. The writers' choice of the star 61 Cygni was obvious: 61 Cygni is a binary star; in 1964 one of its two stars was the only one known to have an extra-solar planet. Now we know that extra-solar planets are common in our galactic neighbourhood. Now we also know that lasers can be used for interstellar communication.

In 1960 Drake made the first real attempt to listen to ETs. In Project Ozma (named after the queen of *The Wizard of Oz*, he aimed a 26-m radio telescope at the stars Tau Ceti and Epsilon Eridani, some 11 light-years away. For two months he listened for radio signals at 1.5 gigahertz, the frequency emitted by hydrogen gas. He chose this frequency because hydrogen is the most common element in the universe. Of course, the search yielded nothing. Since then, more than 100 powerful radio searches have also failed to make any contact. Why have no intelligent radio signals been picked up? Some scientists say that listening to radio signals or sending them might not be the right way to make contact. They suggest that an interstellar laser might be a better communicator.

The American physicist Charles Townes, who was awarded the 1964 Nobel Prize for his hand in the invention of the laser, realised around the time that Drake aimed his telescope at distant stars that ET civilisations could just as easily exploit the optical and infrared portion of the spectrum as the radio portion. Decades passed before laser technology had advanced to the point where powerful lasers capable of sending interstellar messages could be made.

Lasers have two main advantages over the millions of radio channels available for broadcasting: light is easier to focus into a tight beam than radio waves, and it is a better carrier of information. A tight laser beam can be easily focused on a target, and it can transmit a whole encyclopaedia in a second—much better than merely asking 'Is anyone out there?' by a radio wave. But lasers are not as cheap as chips (or radios); they require billions of kilowatts of energy to broadcast for a fraction of a second.

Scientists may one day succeed in sending a laser message to their Cygnian counterparts who mistakenly zapped the beautiful Tunguska taiga. An apology is overdue.

Radioactive Tunguska

In his book, *Siberia: The New Frontier* (1969), the American author George St George, who spent most of his childhood in Siberia, describes his account of travels there in the mid-1960s. In the book, he also makes a brief reference to Tunguska. An excerpt:

> Some investigators seem to believe that whatever flashed across the taiga was intelligently directed because they feel only this explains the changing course of its flight. Was it then some sort of interplanetary vehicle in trouble, perhaps intentionally destroyed by its crew? Quite a few serious scientists seem to believe so, including some Soviet ones like Felix Zigel. Every serious UFO investigation organization throughout the world lists the Tunguska explosion in

connection with possible interplanetary visitors who presumably have visited and studied our planet.

I remember this matter being discussed in our home in Chita, in Transbaikalia, probably in 1914, by my father and his friend, a doctor, who claimed to have visited the site of the Tunguska explosion a few months after it had occurred. The doctor had a detailed diagram showing the zig-zag course of the falling body over some 100 miles (where the tops of trees were sheared off) before the actual explosion. He also said that some unusual glow was observed each night over the epicentre of the explosion for weeks after it had occurred, suggesting some sort of radiation. My father, who was interested in the so-called 'flying saucer' lore even then, was convinced that interplanetary visitors were using some parts of the taiga as their terrestrial base. He drew this conclusion from Evenk legends ... Unfortunately, all my father's voluminous notes on the subject were lost in China where he died in a Buddhist monastery in 1928.

St George's account is intriguing because it mentions a visit to the site by his father's doctor friend in 1914. This is one of the few known accounts of a visit by a non- Evenk person before Kulik's first expedition in 1927. You can draw another conclusion from this account: even in the dark days of the Tsarist empire, Russians were as fascinated by UFOs as Americans are today. St George also notes that 'no dangerous radiation has been found there at present, so perhaps Tunguska will in time become a familiar tourist attraction as the Arizona crater is today'. The phrase 'no dangerous radiation' is interesting, as it reflects the preoccupation of many Soviet scientists in the late 1950s and early 60s with the idea that the Tunguska site was awash with radiation. The main proponent of this idea was the geophysicist Aleksei Zolotov.

Zolotov was as enigmatic as the Tunguska event itself. In a 1978 special programme to celebrate the 70th anniversary of the event, Moscow Radio described him as 'another noted investigator'. However, in its 70th-anniversary report on Tunguska, the journal *Nature* saw him in a different light: 'His name turns up unfailingly in any discussion of the problem, and his theories, however bizarre to the scientific establishment, do at least get published ... his own academic background seems obscure, and according to one physicist who worked for many years on the Tunguska site, Zolotov was originally simply an oil technologist, co-opted on to an expedition for his knowledge of the local terrain!' In the scientific literature of the time, he is referred to as a 'prominent geophysicist'. Ten years later, in 1988, by the time he had led twelve Tunguska expeditions, Nature agreed that Zolotov had 'gradually emerged as an authority in his own right'.

One area of Zolotov's authority was his atomic theory of the Tunguska explosion. He garnished Kazantsev's glass of 'exploding spaceship' vodka with a twist of lemon—the explosion was not an accident. In 1975, when he was head of a Soviet scientific team studying the phenomenon, he suggested that the aliens deliberately detonated the spaceship simply to let us know of their existence. The actual area of destruction was 'an amazing demonstration of pinpoint accuracy and humanitarianism,' he pointed out.

In 1980, the American science writer T.R. LeMair expanded upon Zolotov's 'humanitarianism' idea in his book *Stones from the Stars*. He claimed that the time of the Tunguska explosion seemed 'too fortuitous for an accident'. If the Siberian missile had met Earth just 4 hours and 47 minutes later, it would have scored a bulls-eye hit on the seat of the tsarist empire; and a tiny change of course would have devastated populated areas of China and India. He suggested that 'the flaming object was being expertly navigated' using Lake Baikal as a reference point: 'The body approached from the south, but when about 140 miles from the explosion point, while over Kezhma, it abruptly changed course to the east. Two hundred and fifty miles later, while above Preobrazhenka, it reversed its heading toward the west. It exploded above the taiga.' A thorough scientific review of eyewitness accounts suggests otherwise: the object did not change its course as it moved across the sky from south-southeast to north-northwest.

Zolotov's major contribution to the Tunguska folklore is not in deciding whether the spaceship exploded by accident or by design, but in the radioactivity, it added to the explosion site. He was, in fact, simply cloaking

Kazantsev's science fiction ideas in scientific respectability. While the images of 'the fireball and the mushroom cloud' of an atomic bomb made Kazantsev see a spaceship soaring in the Tunguska sky, the images of people dying with atomic bomb radiation convinced him that the 'survivors' of Tunguska were also exposed to Hiroshima-like radiation doses. 'It could be nothing other than radioactivity,' explains one of the characters in his science fiction novel, *Visitor from the Cosmos*, when a man, shortly after examining the blast area, dies in excruciating pain as if from an invisible fire.

Like Kazantsev, Zolotov had also won Zigel's support. In *Znanyie-Sila* magazine in December 1959, Zigel discussed the results of Zolotov's expeditions of the past three years. During these expeditions, among other things, he compared the effects of the ballistic waves caused by the velocity of the Tunguska body in the atmosphere and the blast waves caused by the explosion itself. Zolotov's study of trees—those that had remained standing and on which the traces of the effects of both waves remained—showed that ballistic waves arrived from the west and broke only small branches, whereas the blast from the north broke larger branches. Zigel estimated the velocity of the body in its final stage of flight to be a relatively low 4,300 km/h; therefore, the explosion was due to the internal energy of the body, not the energy of its motion. He concluded that the blast waves caused most of the devastation.

Zolotov had found trees some 17 km from the blast centre which had been subjected to heat and started to burn. He ruled out a natural forest fire. He said that to start a fire in a living tree, the heat energy must be between 60 and 100 calories per square cm. Similarly, to have caused a sensation of burning in eyewitnesses 70 km away in Vanavara, the energy must have been not less than 0.6 calories per square centimetre. He estimated the heat energy of the explosion to be about 3.5 megatons. As the estimates of the total energy of the blast were also within this range, he reasoned, the blast was nuclear.

Evidence soon started appearing to support Zolotov's popular imagination: mysterious scabs suffered by surviving reindeer (burns from hot ash?), tree rings suggesting enormous growth rate after the blast (normal after wildfires?), high levels of radioactive carbon-14 in the soil and peat collected from the region (not enough to support the idea of a nuclear explosion?), and so on.

Although Hiroshima and Nagasaki showed us the horrors of radioactivity, some radioactivity happens constantly all around us. Small amounts of radioactive atoms are found in the soil we stand on, the food we eat, the water we drink, the air we breathe. This is known as background radiation. Our daily dose of background radiation varies from place to place, but average annual levels typically range from about 1.5 to 3.5 millisieverts (150 to 350 millirems). Eighty per cent of this average comes from natural sources such as indoor radon, food and drink, and rocks and soil. The remaining 20 per cent comes from artificial radiation sources, primarily X-rays. From the study of cancers in survivors of Hiroshima and Nagasaki, scientists have estimated that they were instantaneously exposed to thousands of times the average annual background radiation dose, which continued to increase from long-term fallout.

Most expeditions to Tunguska in the late 1950s and early 60s concentrated on finding the effects of radioactivity on the site. After the 1958 expedition, the Soviet Academy of Sciences decided against any new expedition, but to concentrate on the study of rock and soil samples already collected. This decision led to the formation of the Interdisciplinary Independent Tunguska Expedition (IITE, known as KSE in Russian). The KSE was formed in 1958 in the Siberian city of Tomsk, under the leadership of Gennady Plekhanov, a physician as well as an engineer at the Betatron Laboratory of the Tomsk Medical Institute. KSE was, in fact, formed to discount spaceship theories. The founders even jokingly suggested that 'we must find a nozzle from the spaceship'.

Marek Zbik has described the first KSE expedition in the *Bulletin of the Polish Academy of Sciences*. The expedition, led by Plekhanov, included medical students who collected information to test the hypothesis about post-radiation illness among the Evenk people. 'No trace of such illness was detected,' Zbik writes. 'They also planned to collect bones from the corpses of Evenk people who had died after the catastrophe ... It was not easy to find such corpses because the Evenk people kept the burial sites a secret'. However, students were successful in examining the bones of people who had died during the smallpox epidemic of 1915. The results of these limited investigations did not confirm increased radioactivity in the bones tested.

Plekhanov also collected 300 soil samples and nearly 100 plants. An analysis of these samples in Tomsk showed that 'in the centre of the catastrophe radioactivity is one and a half to two times higher than 30 or 40 km away from the centre'. Plekhanov refused to speculate on the cause of this radioactivity. In another study, he compared the bright nights seen in parts of Europe and Asia after the Tunguska explosion with those following the high-altitude nuclear tests conducted by the United States at Bikini Atoll in 1958. He found that similar atmospheric effects followed both explosions. In short, Plekhanov failed to find any 'spaceship nozzle'. He left the KSE in the 1963 but remained active in research on Tunguska

Kirill P. Florenskiy and Vassilii Fesenkov, two of the main proponents of the comet theory in the 1960s, violently opposed Zolotov's nuclear explosion ideas. 'There are no planets with a highly organized life from which such a ship could descend,' Fesenkov told *The New York Times* in 1960. 'This suggestion has now been rejected by most of the Soviet scientific community,' the *Times* added.

Florenskiy devoted most of his time on the 1961-62 expedition to trying to disprove Zolotov. He concluded that the radioactivity at the centre of the blast was within the range of fluctuations of the present background radiation. However, he agreed that it was somewhat higher at the centre than it was a few kilometres away. Most of the radioactive atoms were concentrated in the upper layers of soil and peat. He suggested that they had been accumulated from global fallout from atomic and hydrogen bomb tests. On the matter of accelerated tree growth in the devastated area, Florenskiy said that the growth was not due to genetic mutation from radiation but was 'only the normal acceleration of second growth after fires had taken place'.

More recently, in 2001 Academician Nikolai Vasilyev said: 'The results of a search for radioactivity in the region of the Tunguska explosion negate a nuclear hypothesis. It should be noted, however, that a search for traces of radioactivity fallout half a century after a nuclear explosion in the atmosphere is a challenging task, especially considering contamination from recent atmospheric nuclear tests.'

The Russian scientists Victor Zhuravlev and A.N. Dmitriev have developed a plasmoid hypothesis for the Tunguska body: a sort of bottle filled with plasma and surrounded by a strong magnetic field. This 100,000-tonne plasmoid was ejected from Sun. Vitalii Bronshten, the major proponent of the modern comet theory, who died in 2004, was highly critical of this attempt 'to disguise a spaceship as a plasma container'. 'This is a typical ad hoc hypothesis,' he said. 'We use this example to demonstrate that all ad hoc hypotheses, a great number of which have been proposed to explain the Tunguska event, are useless.'

Zhuravlev disagrees. Three graphs of magnetic activity found in 1959 at the Irkutsk Magnetic and Meteorological Observatory show a magnetic storm that started soon after the Tunguska event and lasted for about four hours. The magnetograms have nothing in common with those caused by meteorites but have all the distinctive features of the disturbances of the geomagnetic field that are generated by nuclear bombs. The Tunguska object, says Zhuravlev, was 'a cosmic object the composition and structure of which is unknown to astronomers and physicists'.

The images of a nuclear spaceship exploding above the Tunguska taiga remain vivid in many researchers' minds in the 21st century. One of them is Vladimir V. Rubtsov of the Research Institute on Anomalous Phenomena in Ukraine. His 'ad hoc hypothesis' is the so-called battle model: in 1908 there was an aerospace battle between two alien spaceships, after which one of them survived and flew back to space. 'Perhaps one day in the future it will be possible to deduce a convincing model of the phenomenon directly from facts accumulated,' he writes in a newsletter of the Institute.

Another is Yuri Lavbin of the Tunguska Spatial Phenomenon

Foundation in Krasnoyarsk, which is a group of physicists, geologists and mineralogists who have been organising regular expeditions to the explosion site since 1994. Lavbin believes that the explosion was caused by the collision of an extraterrestrial spaceship with a comet.

In the summer of 2004 Lavbin announced that his team had found two strange black metallic blocks near the site. The 50-kg blocks, Lavbin claimed, are the remnants of a spaceship. 'Their material recalls an alloy used to make space rockets, while at the beginning of the 20th century only planes made of plywood existed,' he said. The meteorite committee of the Russian Academy of Sciences has dismissed Lavbin's claim saying that in Siberia where oil geologists regularly work 'you can find a heap of fragments of various machines.'

Spaceship Tunguska still soars in the Siberian sky. Did Tesla's death ray fail to zap it?

Chapter 10

Tesla's Death Ray

Nikola Tesla was a genius so ahead of his time that his contemporaries failed to understand his ground-breaking inventions. 'If ever an inventor satisfied the romantic requirements of a Jules Verne novel it was Tesla,' said a *New York Times* editorial on 9 January 1943, after his death. 'If that abused word "genius" ever was applicable to any man it was to him.'

He was 'a first-class mathematician and physicist whose blueprints were plausible, even though they were far ahead of the technical resources of his day,' the *Times* editorial added. He was so much misunderstood as a great scientist that he became the inspiration for the mad scientist in Max Fletcher's *Superman* cartoons of the 1940s.

An inventor of dazzling brilliance who belonged to the 20th century's heroic age of invention, of which Edison was the most distinguished exemplar, he invented and developed AC power, induction motors, dynamos, transformers, condensers, bladeless turbines, mechanical rev counters, automobile speedometers, gas-discharge lamps that were the forerunners of fluorescent lights, radio broadcasting, and hundreds of other things (the number of patents in his name exceeds 700).

An eccentric who preferred science to society, he became a virtual recluse for the last quarter-century of his life. He never married, never developed any close relationship. 'He made everybody keep at a distance greater than three feet,' according to the manager of a New York hotel where he had spent the last years of his life in the company of his pet pigeons. His strategy of keeping a safe 'social distance' would have served him well in the 2020 coronavirus pandemic. A man ahead of his time about his knowledge of electricity and germs, indeed.

Tesla, as a 1999 biography claims, was the man who invented the 20th century, although he was almost forgotten after his death in 1943. But not forever.

On the centenary of his birth in 1956, he was honoured by scientists when they named a unit for measuring magnetism (the SI unit of magnetic flux density, the tesla) after him. This 'nominal immortality' has placed him in the company of Ampère, Volta, Ohm, Gilbert, Henry, Faraday and Hertz, great scientists who have all had electromagnetic units named after them.

Five decades after receiving recognition from his peers, he has now also been given 'cyberspace immortality' from adoring fans in countless web pages of biographies, essays, 'my science hero' projects, online museums, discussion groups, and so on. If the number of web pages can be considered a measure of public popularity, Tesla is now catching up with Marconi but is still a long way away from Edison—two contemporary inventors who have become legends. A surf through Tesla web pages gives the impression that the enigmatic inventor has become a cult hero and has found a place in the hearts of the fans of UFOs, free-energy generators, anti-gravity machines and such other alternative science ideas. Numerous web pages are devoted to his death ray, an invention that links Tesla to Tunguska.

Tesla was born at Smiljan, Croatia (then part of Austria-Hungary) on 10 July 1856. His father was a Greek clergyman and orator, and his mother an inventor of home and farm appliances. After graduating from a high school in Carlstadt, Croatia, he studied engineering at the University of Graz. In 1884 he emigrated to America.

When the 28-year-old arrived in New York, he had four cents in his pocket and a few papers in his suitcase which were scribbled with a drawing and some mathematical calculations for an idea for a flying machine. He lived and worked in New York for almost 60 years. When he died in his hotel room on 7 January 1943, he was penniless. Still, his room was full of scientific papers and plans so revolutionary that some of them are rumoured to be the blueprints for a missile defence system similar to the US Strategic Defense Initiative (popularly known as 'Star Wars' missile defence) of the 1980s.

His 'practical inventions' were limited to a short period from 1886 to 1903. It was the Jules Verne future that engrossed him, according to the *Times* editorial: 'Communicating with Mars, plucking heat units out of the atmosphere to run engines, using the whole earth as an electrical resonator so that a man in China could communicate wirelessly with another in South America, transmitting power through space—it was to such possibilities that he devoted the last forty years of his long life.'

In later years of his life, Tesla was a favourite of newspaper reporters who revelled in recounting his incredible inventions. On his 78th birthday, Tesla told a *New York Times* reporter that he had invented a death ray powerful enough to annihilate an army of 10,000 planes and one million soldiers instantaneously. The next day, 11 July 1934, the paper ran a story which was headlined in the style of the time:

TESLA, AT 78, BARES NEW 'DEATH-BEAM'
Invention Powerful Enough to Destroy
10,000 Planes 250 Miles Away, He Asserts.
DEFENSIVE WEAPON ONLY
Scientist, in Interview, Tells of Apparatus
That He Says Will Kill Without Trace.

The story referred to Tesla as 'the father of modern methods of generation and distribution of electrical energy,' and quoted him as saying that this latest invention of his would make war impossible: 'It will be invisible and will leave no marks behind it beyond the evidence of destruction. This death-beam would surround each country like an invisible Chinese Wall, only a million times more impenetrable. It would make every nation impregnable against attack by airplanes or by large invading armies.'

On his 84th birthday, Tesla declared that he stood ready to divulge to the United States government the secret of his 'teleforce', with which aeroplane motors would be melted at a distance of 400 km so that an invisible wall of defence would be built around the country. He said that this teleforce was based on an entirely new principle of physics that 'no one has ever dreamed about,' and would operate through a beam one-hundred-millionth of a square centimetre in diameter. The voltage required to produce this beam would be about 50 million volts, and this enormous voltage would catapult microscopic electrical particles of matter on their mission of defensive destruction, he added.

Tesla probably conceived the idea for his death ray at Wardenclyffe, Long Island, New York, where in 1902 he built a 57-m tower and laboratories to experiment on radio waves and on transmitting electrical power without wires. The tower's steel shaft ran 36 m underground, and it was topped with a 55-tonne, 20-m diameter metal dome. This experimental facility had the financial backing of the legendary investor J. Pierpont Morgan. However, Morgan pulled out of the venture even before construction was complete. The tower was abandoned in 1911 and demolished in 1917. The main building still stands today.

The popular story that Tesla tested his death ray one night in 1908 goes something like this. In 1908, Arctic explorer Robert Peary was making the second attempt to reach the North Pole, and Tesla requested him to look out for unusual activity. On the evening of 30 June, accompanied by his associate George Scherff atop the Wardenclyffe tower, Tesla aimed his death ray towards the Arctic, to a spot west of the Peary expedition. Tesla then scanned the newspapers and sent telegrams to Peary to confirm the effects of his death ray but heard of nothing unusual in the Arctic. When Tesla heard of the Tunguska explosion, he was thankful no one was killed, and dismantled his death ray machine, feeling it was too dangerous to keep it.

In a letter to *The New York Times* on 21 April 1907, Tesla wrote: 'When I spoke of future warfare, I meant that it would be conducted by direct application of electrical waves without the use of aerial engines or other implementation of destruction ... This is not a dream. Even now wireless power plants could be constructed by which any region of the globe might be rendered uninhabitable without subjecting the population of other parts to serious danger or inconvenience.'

Though he believed that it was 'perfectly practicable to transmit electrical energy without wires and produce destructive effects at a distance', there is no evidence that Tesla used the Wardenclyffe tower for his experiments on the death ray.

In an interview with authors Walter W. Massie and Charles R. Underhill for their book *Wireless Telegraphy and Telephony* (1908), Tesla explained his vision of the future uses of radio waves:

'[My experiments] will make it possible for a businessman in New York to dictate instructions and have them instantly appear in type at his office in London or elsewhere. He will be able to call up, from his desk, and talk to any telephone subscriber on the globe, without any change whatever in the existing equipment. An inexpensive instrument, not bigger than a watch, will enable its bearer to hear anywhere, on sea or land, music or song, the speech of a political leader, the address of an eminent man of science, or the sermon of an eloquent clergyman, delivered in some other place, however distant. In the same manner, any picture, character, drawing, or print can be transferred from one to another place. Millions of such instruments can be operated from but one plant of this kind.'

He added that he considered the transmission of power without wires more important than his work on radio waves and that his experiments would show it 'on a scale large enough to carry conviction'. He never realised his vision for the Wardenclyffe tower because of the lack of financial backers. On 17 February 1905, he wrote to Morgan pleading again for the help: 'Let me tell you once more. I have perfected the greatest invention of all time—the transmission of electrical energy without wires to any distance, a work which has consumed my life'.

There is no reference in his Wardenclyffe tower works to how his system would work. The only reference is in a highly technical article, 'The New Art of Projecting Concentrated Non-Dispersive Energy through the

Natural Media,' written in 1937. In this article, he described the actual workings of a particle-beam weapon for destroying tanks.

The death ray may have been a plausible dream, but it was not a reality. Tesla never got the opportunity to test his plans. The Tunguska story seems improbable for another reason. Tesla could not have heard about the Tunguska event before 1928 when stories about it appeared in the American newspapers. Also, there is no record of Tesla's request in Peary's accounts of his expedition. The story has simply been conjured up by joining the dots—Tunguska, Tesla, Peary—with 1908. But the dots do not interrelate.

The American writer Oliver Nichelson, whose name pops up in many books and web pages that link Tesla with Tunguska, believes that the idea of a Tesla-directed energy weapon causing the Tunguska explosion was incorporated in a 1994 fictional biography by another writer, and was the subject of a segment on the TV documentary series *Sightings*. 'Given Tesla's general pacifistic nature it is hard to understand why he would carry out a test harmful to both animals and the people who herded the animals even when he was in the grip of financial desperation,' he says. 'The answer is that he probably intended no harm, but was aiming for a publicity coup and, literally, missed his target.'

The evidence is only circumstantial, Nichelson agrees. However, he still wants to bet on both heads and tails: 'Maybe the atomic bomb size explosion in Siberia near the turn of the century was the result of a meteorite nobody saw fall. Or, perhaps, Nikola Tesla did shake the world in a way that has been kept secret for more than a century'. You spin the coin.

Chapter 11

New Fireballs in the Siberian Sky

Ninety-four years after the Tunguska fireball. About 11.50 p.m., 24 September 2002. A remote, semi-mountainous and sparsely populated region of Siberia near the Vitim River, northeast of Irkutsk and Lake Baikal.

A US Air Force satellite spots an object as it enters the atmosphere but loses track of it as it falls below 30 km. Moments later a second satellite records a fireball exploding in the clouded sky. A white, bright luminescence appears in the southwest and floods the whole sky—the colour of luminescence changes from white to blue and reddish-brown as it disappears in the northeast.

Virtually no one in Siberia sees the fireball, but residents hear the sounds of the explosion up to 60 km from the site. The blast waves of the explosion—with the energy of a small atomic bomb—flatten 100 square km of the taiga, but no one is hurt or killed. The explosion also sends seismic waves which make windowpanes rattle, and house lights swing in the village of Mama 60 km away. For several days residents notice sporadic flashing lights in the direction of the explosion.

A mini-Tunguska

A small team of researchers and journalists from Irkutsk reached the remote area in late October 2002. But despite using satellite data to locate its position, they were unable to identify and reach the impact site (latitude 58 degrees 9 minutes north, longitude 113 degrees 21 minutes east). The researchers collected 25 eyewitness accounts, which generally agreed that 'a large rock fell from the sky and then Earth trembled'. Some residents as far away as 70 km from impact site said that they saw a 'sphere with a tail'. Witnesses also 'heard a roar and splashes of light above the taiga far away'.

Alexander Doroshock, a gold miner, said that 'suddenly the sky turned turquoise, there was a large flash followed by an explosion that produced a sharp whistling sound'. Several other witnesses also talked about hearing rustling and buzzing sounds as the fireball streaked across the sky. These were probably electrophonic sounds, which result from the light given off by a meteorite.

Two Mama airport employees, Vera Semenova and Lidiya Berezan recalled a scary phenomenon: a bright glow at the upper ends of the thin little wooden poles of the fence surrounding the airport's meteorological station. According to Sergei Yazev, the leader of the expedition, the glow was probably caused by a strong electric current produced by the meteorite. As heavy snowfall prevented further searches, the expedition returned to Irkutsk without determining the exact location of the impact site.

In May 2003 an expedition mounted by Kosmopoisk, an organisation of amateur enthusiasts interested in research on various anomalous phenomena, reached the impact site and found an area of about 100 square km covered with fallen trees. Some trees in the centre of this area, where the blast wave touched the ground, showed signs of burning. A few kilometres from this fallen forest, they also discovered the impact site, which was covered in about 20 small craters, up to 20 m in diameter.

On the day after the impact, medical workers in Mama measured radioactivity in local villagers. It showed a two-fold increase above the background radiation but returned to normal within a few days. Medical workers also told the Kosmopoisk group that the health of residents worsened for some time after the event. Water and snow in the blast area also tasted bitter. An analysis of water samples in Moscow showed that there were large amounts of tritium—radioactive hydrogen—in the water, 'like in the water of nuclear power station cooling ponds'.

The Kosmopoisk group failed to find any meteorite fragments, and the members believe that the object was probably a small comet with a weight of about 100 tonnes which broke up in mid-air. 'The character of the damage and the radiation background at the explosion epicentre are substantially different from the after-effects of a meteorite fall,' says the expedition leader Vadim Chernobrov. However, he adds, the discovery of tritium in water is rather strange for comets.

Andrei Ol'khovatov believes that the tritium came from seven underground nuclear tests conducted from 1976 to 1987 about 400 km north of the impact site. 'Siberia is a mysterious place, and you could find almost anything you want there, and human activity could make it even more mysterious,' says Ol'khovatov.

According to Chernobrov, the members of the expedition do not rule out other rare natural phenomena such as giant ball lightning or ejection of minerals from the ground that disintegrated into water and gas. UFO enthusiasts are not ruling out any cosmic visitor either.

When reporting news of the second-largest meteorite, after the Tunguska meteorite, to fall in Siberia within a century, newspapers around the world were more interested in speculating what would have happened if the fireball had exploded over a big city.

'If it hit Central London, Britain would no longer have a capital city,' speculated the London Times.

'It was an explosion of such a force that if the supposed fireball had fallen on Moscow, half of the Russian capital would have turned into desert, and the other half into ruins,' echoed the Russian news agency Novosti.

The news of the Vitim fireball has also made asteroid doomsayers worried because such an impact would force them to revise their estimate for the likelihood of a devastating asteroid striking Earth (meaning: Asteroid Armageddon is closer than you think.)

More trouble from space

The news of another rogue asteroid came on 15 February 2013 when a 20-m-wide asteroid travelling at a speed of 540 km/h exploded over the city of Chelyabinsk, Russia. About 2,000 km west of the Tunguska blast site, it was the largest impact event on Earth since 1908. Thousands of people saw a bright flash in the sky followed by a powerful sonic boom.

The shock wave it generated damaged thousands of homes and injured more than 1,500 people, mainly broken glass, in and around Chelyabinsk. According to an Associated Press report, lessons had just started at Chelyabinsk schools when the meteor exploded, and officials said 258 children were among those injured. 'A resident, Valya Kazakov, said some elderly women in his neighbourhood started crying out that the world was ending,' Associated Press reported.

The explosion spawned tremendous sonic booms, bright lights and chunks of rocks falling from the sky. The largest fragment, a whopper rock probably weighing half a tonne, may have splashed down in Lake Chebarkul, 70 km west of Chelyabinsk. Its tell-tale sign was a 6-m-wide hole found in the lake's frozen surface the morning after the impact. The blow-up of this 10,000-tonne rock at the height of nearly 32 km scattered smaller rocks weighing several kilograms over a 25-km swathe south of the final point of the path of the asteroid.

Scientists predict a Tunguska-like event once every millennium, and a Chelyabinsk-like event once a century. Coincidentally, the Chelyabinsk explosion happened on the same day an asteroid, called 2012 DA14, was flying by Earth. Just hours later, it passed within 27,000 km of Earth. It was travelling in a direction that was opposite to that of the Chelyabinsk asteroid. At the moment, no country has a system to shoot down objects falling from the sky.

You have not yet heard the last word on rogue asteroids.

Chapter 12
Tunguska Revisited

In 1972, John Anfinogenov, a Russian scientist, found an exotic quarzitic boulder entirely buried in soil and permafrost close to the epicentre of the Tunguska fireball.

Roughly almond in shape, about 2 m x 1.5 m x 1 m in size, the boulder is estimated to weigh about 10,000 kg. It is made almost entirely of quartz grains of about 100 micrometres in size. The boulder is now known as:

John's Stone

In 2014 and 2015, writing in the journal *Icarus*, John Anfinogenov and his colleagues presented evidence to show that John's Stone is a possible fragment of the Tunguska meteorite and that the trajectories of John's Stone landing and Tunguska body flights coincide.

They said that the following findings confirm the high-speed movement of John's Stone through the near-surface of permafrost layers: (a) buried slightly arching furrow in the permafrost; (b) shear-fractured splinters of the boulder; (c) the boulder's own furrow; (d) splinters with glassy coatings; (e) half-cleaved pebbles on the boulder and its splinters; (f) clear consistency in geometry of spatial arrangements of all splinters, furrows, cleaved pebbles etc.; and (g) and landing velocity of the boulder.

Excavated trenches around the boulder reveal fresh disturbances in permafrost that suggest that the boulder impacted the ground from above. Quartz is extremely rare in meteorites, and it may have been ejected from Mars as quartz or silica-rich rocks appear to be present on Mars

Leonid Kulik, the first Tunguska investigator who devoted the last two decades of his life to searching for a meteorite fragment, would be delighted to learn in his grave that Russian scientists have finally found one. At the same time, he would be disappointed to learn that the evidence is not yet conclusive.

Muddling the image

In 2012, after a detailed reconstruction of the Tunguska phenomenon, mathematician Vladimir Rubtsov of the International Institute of Safe Technologies at Kharkov in Ukraine concluded that none of the existing hypotheses fits the image sufficiently well. In a paper posted on arxiv.org, he said that the high energy concentration of the Tunguska explosion contradicted the hypothesis of the vapour cloud explosion. And an ordinary comet or a stony asteroid seemed out of the question.

What was it then? 'At present, we do not know,' he writes. But we know that instead of elements that are present in space—iron, nickel and cobalt, it contained titanium, aluminium, ytterbium and gold. The Tunguska space body flew at a low velocity and exploded due to the internal energy of its substance, not due to the energy of motion ... So, if the Tunguska space body was a natural space body, then it means there exists in space another type of dangerous space objects, whose nature remains vague at best.'

NASA seems to disagree.

NASA's new research

Every day, many tonnes of tiny rocks—smaller than pebbles—hit Earth's atmosphere and disintegrate. But mid-size rocks that can go through the atmosphere and do severe damage are rare. The new research from NASA shows that the impact of such mid-size rocks is less frequent than previously thought. Such devastating impacts happen in the order of millenniums, not centuries, as previously thought.

In 2018, at a workshop at NASA's Ames Research Center, about 50 experts on impact airburst re-examined the Tunguska event and concluded that it was a stony (not icy) body between 164 and 262 feet (50 to 78 metres) in diameter. It entered the atmosphere at around 34,000 miles (per hour 54,700 km) per hour and exploded at an altitude of 6 to 9 miles (9.6 to 14.5 km). The explosion released energy in the range of 10 to 20 megatons (Mt).

According to David Morrison, a planetary science researcher at Ames: 'Tunguska is the largest cosmic impact witnessed by modern humans. It also is characteristic of the sort of impact we are likely to have to protect against in the future.'

On 15 February 2013, a smaller but still impressive fireball exploded near Chelyabinsk, Russia. The current understanding of the asteroid populations suggests that an object like the Chelyabinsk meteor can impact Earth every 10 to 100 years.

Why must we find the answer?

The Tunguska explosion was a cataclysm that has happened countless times in Earth's history, and it is sure to happen again—that's what Academician Nikolai Vasilyev (1930-2001) believed. 'Had such a cosmic body exploded over Europe instead of the desolate region of Siberia, the number of human victims would have been 500,000 or more, not to mention the ensuing ecological catastrophe,' he said. 'The Tunguska episode marks the only event in the history of civilization when Earth has collided with a truly large celestial object, although innumerable such collisions have occurred in the geological past. And many more are bound to occur.'

He stressed that continued investigations of the Tunguska event were necessary, simply because it would happen again. Only by knowing what the object was, and by recognising its devastating biological consequences, will the scientific and medical communities be in a position to deal with such a cataclysm in the future. 'Today the Tunguska problem can be considered as an important part of the larger problem of a possible collision of Earth with near-Earth asteroids,' Vasilyev said. The most important question for us is: can it happen again? We'll never know the answer.

Like the riddle of the death of the dinosaurs, the Tunguska mystery still eludes scientists. Many believe that this cosmic mystery would finally be solved only if we found ponderable fragments of the object. At least, let's try to guess whodunit.

Chapter 13

Whodunit?

It is surprising that, after eleven decades, international scientists remain divided. Chris Trayner of Leeds University seems to have solved the riddle of why scientists have failed to solve the mystery of Tunguska. He says that many of their problems stem from the old East-West divide and poor access to Russian literature. He suggests that to disentangle what might have happened in 1908, scientists need extensive translations and online indexes of the Russian source material. The same must be said of English sources for Russian scientists.

In the meantime, let's try to disentangle what really happened. There are seven primary sources of information on the event: (1) the devastated forest and its pattern of damage (first observed nineteen years after the event); (2) records of atmospheric and seismic waves at the time of the event; (3) records of magnetic storms at the time of the event; (4) bright nights observed in parts of Europe and Asia after the event; (5) unusual atmospheric phenomena seen after the event; (6) study of microscopic particles found at the explosion site and in Antarctica; and (7) eyewitness accounts (first collected 13 years after the event).

Almost everyone agrees on the following points:

- The accurate time of the event (0014 GMT; 7.14 a.m. local time) and the exact location of the epicentre (latitude 60 degrees 55 minutes north, longitude 101 degrees 57 minutes east).
- A large incoming object, presumably a meteorite or a comet, was seen over an area 1,500 km across. The object's brightness was comparable to that of the noonday sun.
- It exploded in mid-air, between 5 and 10 km above the ground. The energy of the blast was probably between 10 and 20

megatons.
- The debris was blown upwards into the stratosphere. No significant fragments, except microscopic globules at the epicentre and in Antarctica, have been found.
- The blast created a shock wave that levelled 2,150 square km of forest, of which a heatwave had burnt about 200 square km. In a forest fire, trees are usually burnt on the lower part of their trunks, but these trees had been burnt uniformly and continuously.
- There is no impact crater at the epicentre, the point under the explosion where the shock wave first contacted the ground.
- The devastated forest has the shape of a butterfly spread over the land. The epicentre lies near the 'head' of the butterfly.
- The heat and shock waves were felt by many eyewitnesses at Vanavara, about 70 km from the explosion site. Bright lights in the sky were seen as far as 700 km away; and loud explosions, like gunfire, were heard after the explosion up to 1,200 km away.
- The blast also created disturbances in the atmosphere which were recorded around the world. The impact of the blast on the ground generated seismic waves which were recorded well beyond Russia.
- A local magnetic storm—similar to ones produced by nuclear explosions in the atmosphere—began a few minutes after the blast and lasted for about four hours.
- The explosion caused very bright nights in parts of Europe and Asia which lasted for several nights. Noctilucent clouds were also observed. The dust in the stratosphere generated these optical anomalies.
- There was accelerated growth in young trees that survived the blast.

The controversial points:

- The object's shape as it raced across the sky before exploding: a 'pipe,' 'pillar' or 'tube,' as mentioned by some eyewitnesses? Did it

leave a trail of smoke and dust?
- The object's nature, size and mass; for how long it was seen in the sky.
- The object's entry angle, its subsequent flight path, and its velocity immediately before the explosion.
- The duration of the explosion.
- The flight ending in multiple explosions.
- Increased radioactivity at the explosion site; genetic effects on the local population; the accelerated growth of trees being caused by a mutation.
- Craters such as the Suslov crater being formed by the impact of the exploding object.
- Optical anomalies being observed before the event as early as 23 June.

Bordering on the ridiculous:

- The object twice changed course in flight.

The usual suspects

As far back as 1969, a Russian writer, P.I. Privalova (believed to be the pseudonym of Igor Zotkin, a member of the then Committee of Meteorites of the Soviet Academy of Science), published a list of 77 theories that had been put forward to explain the Tunguska event. The list could be stretched to 120, Ms Privalova hinted, if one was at a campfire in the taiga with a glass of vodka. As the following list has been compiled without the benefit of a glass of vodka at a taiga campfire, it includes only a 'dirty dozen'.

A comet that disintegrated in the atmosphere. The unusually loose structure of the comet led to its break-up in the atmosphere. The diameter of its nucleus is estimated to be 40 m, much smaller than diameters of visual comets.

An asteroid that exploded in mid-air. A stony asteroid that exploded at about 8 km above the ground. It was about 30 m across. The 15-megaton explosion released a million tonnes of small particles in Earth's atmosphere. Winds dispersed the dust in the stratosphere, which caused the bright skies reported in the aftermath of the fall.

A mini black hole that passed through Earth. A mini black hole, invisible to the naked eye, entered at Tunguska and then travelled through Earth for about 15 minutes. When it exited through the North Atlantic, it caused shock waves in the ocean and the atmosphere.

An antimatter rock that annihilated itself when it entered the ordinary matter atmosphere. The explosion, which was like a 35-megaton atomic or hydrogen bomb, generated trillions of radioactive carbon-14 atoms.

A mirror matter rock that nobody could see. As it dived into the atmosphere, the heat caused it to explode at high altitude. The rock was roughly 100 m in size and weighed about one million tonnes.

A volcanic blow-out. Natural gas escaped from narrow underground volcanic vents and raced upwards at high speed and started mixing with air. After a few hours this volatile mixture, which contained 10 million tonnes of methane, exploded like a fireball.

A giant lightning ball that materialised from nowhere. The estimates of its diameter vary from 200 m to one kilometre. It disintegrated into smaller spheres, which further disintegrated into still smaller spheres, until finally, they exploded.

'Geometeors' that came from below. The explosion was caused by a strong coupling between some unknown subterranean and atmospheric processes. This coupling formed meteor-like luminous objects, but of terrestrial origin.

A plasmoid surrounded by a strong magnetic field. This 100,000-tonne plasmoid was ejected from Sun to wreak havoc at Tunguska.

An alien spaceship that broke down. It plunged out of control through Earth's atmosphere, and within a fraction of a second, it vaporised in a blinding flash of light.

A zap from an alien laser. A laser sent by ETs from a giant planet orbiting the star 61 Cygni zapped the taiga.

An experiment on a 'death ray' which got out of hand. Nikola Tesla mistakenly pointed his death ray at Tunguska.

In the witness box

A tiny comet with a long tail. The available evidence on the object's orbit is consistent with the orbits of Earth-crossing asteroids, but not with the orbits of short-period comets. The collision of comet Shoemaker-Levy 9 with Jupiter has shown us that the mass of a comet that enters a planet's atmosphere needs to be more than 100 million tonnes to trigger a powerful explosion at the end of its journey. The Tunguska fireball's pre-explosion mass is believed to be one million tonnes. It is inconceivable that such a small comet could survive the intense atmospheric pressure on its journey to the Siberian sky. On the other hand, an asteroid could survive such an arduous journey. It has to be an asteroid, a stony asteroid.

A stony asteroid. The facts that prove that the colliding object was a comet include its unusually loose structure, which led to its break-up in the atmosphere; its dust tail, pointing away from Sun, which caused unusual sunsets; and the nature of its orbit. The Sun's glare prevented any sighting of the comet before it hit Earth because its direction and the angle of strike towards Earth were from behind the Sun. It was not an asteroid for another reason: the absence of reasonably large asteroid fragments. No fragments, no asteroids. It was undoubtedly a comet, a ball of ice and dust.

A mini black hole. It's not theoretically impossible that a mini black hole could pass through Earth, but the black hole theory was a good try by a couple of theoretical physicists. It flopped when no one could find any record of a black hole's exit on the other side of Earth through the North Atlantic. Surely it was a big rock, not an invisible black hole.

An antimatter rock. The trouble with these rocks is that they can't survive the journey through the atmosphere. Antimatter-matter annihilation also annihilates this theory.

An invisible mirror matter rock. An excellent effort to test drive the new mirror matter theories. The effect on Tunguska, however, has been hidden. It wasn't a rough rock; it was a beautiful ball.

A dark matter ball. The ball was less than a centimetre wide, but it weighed 100,000 tonnes. It hit the ground with a speed of about 160 km/ and penetrated deep into Earth and only stopped after a few thousand kilometres.

A lightning ball that turned into a killer. No one has yet explained how a tiny lightning ball could turn into a giant fireball. But it's easy for a cloud of methane.

A volcanic blow-out. The radial pattern of fallen trees shows that the shock waves came from above, not from the rattling of Earth. What about the fireball that eyewitnesses saw streaking across Sunny Siberian sky?

Try a geometeor (but the same arguments can be applied against that theory).

Ghostly geometeors. Meteors that do not come from the sky, but the ground under your feet? A plasmoid sounds more plausible.

A plasmoid. A sort of bottle filled with plasma and surrounded by a strong magnetic field? A spaceship disguised as a plasma container? Why not try the real thing?

An alien spaceship: Was it on its way to the Star Wars studios and got caught in a space-time vortex which spewed it onto the set of *The X-Files: Tunguska*? A laser beam is a better bet.

A sharp laser beam. You are not alone, Earthlings. Read your next postcard from space carefully; otherwise, you'll be zapped by a death ray.

A deadly death ray. A yarn that seems to have been spun on April Fools' Day. Our benevolent Dr Tesla fired his death ray to deflect the asteroid that was on its way to destroy his home city of New York. The death ray changed the course of the rogue rock, and it exploded in almost empty Siberia. Thank you, Dr Tesla.

The experts' testimony

Academician Nikolai Vasilyev: 'Though the final choice between the asteroid and the comet theories has not yet been made, the chances of the stony asteroid version have recently grown substantially … The comet theory has lost its dominating position, but it is not clear whether the pressing problems can be solved within the asteroid alternative.'

Dr Vitalii Bronshten: 'We astronomers know two types of solar system bodies—comets and asteroids. We do not know anything else'. (A remark made to Dr Ol'khovatov about the geometeor theory.)

Professor Wolfgang Kundt: Some scientists have thought about this mystery outside the square (or the elliptical orbit of an asteroid or a comet). Kundt is one of them. After analysing statistics of extraterrestrial and terrestrial events, he concluded that tectonic events such as earthquakes and volcanoes are 20 times more frequent than asteroid or comet impacts at the same destruction energy. He says that the disappearance of Sodom and Gomorrah in biblical times may also be due to a volcanic eruption, making the two cities slide to the bottom of the Dead Sea.

His arguments for a tectonic explanation for the Tunguska event—a volcanic blow-out in which millions of tonnes of methane gas escaped from a volcanic vent and turned into a fireball—are entirely plausible. When the gas was suddenly released, it rose, burned and formed clouds in the upper atmosphere for several days, where they scattered sunlight. The scattered sunlight at night gave rise to the bright nights of Tunguska.

'We live on a tectonically active planet,' he says. 'To me, the case of Tunguska is crystal clear, multiply redundant'. There's just one hitch: Kundt's explanation fails to account for the fireball that was seen by many witnesses racing across the sky.

The conclusion of a 'Revisiting Tunguska' workshop held in 2018 at NASA's Ames Research Center: It was a stony (not icy) body between 50 to 80 metres in diameter. It entered the atmosphere at around 54,700 km per hour and exploded at an altitude of 9.6 to 14.5 km. The explosion released energy in the range of 10 to 20 megatons.

The verdict

It seems that the jury is still out.

What would be the verdict of a dispassionate science writer who has no pet theory to push and who has diligently investigated this cosmic whodunit?

'I'll be judge, I'll be jury,' said cunning old Fury (in *Alice's Adventures in Wonderland*). 'I'll try the whole cause and condemn you to death.'

No, I won't do that. But first, a trip back in time.

About 65 million years ago

The tectonic plates that carry the continents and oceans on their backs like giant rafts continue to drift. The continents are flatter, mountain ranges lower, and sea levels higher. Antarctica and Australia are connected. India is an island in the Indian Ocean drifting northwards towards Asia. Ocean waters flood many sections of the continents. A vast inland sea stretches from the Arctic Ocean to the Gulf of Mexico dividing North America into the Rocky Mountains and high plains regions.

There are no ice caps. The climate is warmer; sea surface temperatures average about 19°C.

Redwoods, cypresses, firs, magnolias, oaks, sassafras, ferns and flowering shrubs fill the landscape; grasses have yet to arrive. Long-necked plesiosaurs, giant snake-like mosasaurs, sharks, sea urchin and sea stars inhabit the shallow seas that spill into the continents. Frogs, salamanders, turtles, crocodiles and snake proliferate on the expanded coasts. Giant pterosaurs that soar in the sky face stiff competition from increasingly diverse species of birds. Several hundred types of dinosaurs roam the land from Asia to North America, which are joined by a bridge of land across what is today the Bering Strait. The carnivore *Tyrannosaurus rex*—weighing as much as 9,500 kg, 12 m long and 4 m high—rules this rich and diverse world.

A 10-km-wide chunk of extraterrestrial rock suddenly appears in the sky and hits the shallow sea, near Mexico's Yucatan peninsula, with an unimaginable force. The impact produces an explosion equivalent to 100 trillion tons of TNT—80 billion Hiroshima bombs.

Within seconds a 150-km plume of dust and steam pushes its way into space and expands. After 45 minutes, it envelops the planet. As the plum containing tens of thousands of cubic kilometres of dust falls back to the ground, it creates meteor-like streaks in the atmosphere. These streaks lit the sky like trillions of meteors and heat the atmosphere to several hundred degrees.

The force of the impact also creates an enormous fireball that spreads out, igniting forest fires from North America to Asia. Wildfires ravage the landscape and create trillions of kilograms of greenhouse gases—carbon dioxide, carbon monoxide and methane. The fires die in the long winter after the firestorms.

A week after the impact soot from the fires and dust settles to the ground. The soot covers the planet like a blanket, blocking sunlight for months. Without sunlight, plants stop making food and food chains everywhere collapse. Life on the planet is almost wiped out. So are the dinosaurs, our lovable leapin' lizards.

Fast forward to 1940s

An oil company drilling off the coast of Mexico's Yucatan peninsula stumbles across a tremendous near-circular structure, buried under the surface near the village of Chicxulub (in ancient Mayan, Chicxulub means 'tail of the devil').

Scientific investigations have now revealed that the bowl-shaped structure is indeed an impact crater rather than some volcanic structure. The ring sizes of this multi-ring crater, which is buried beneath many kilometres of carbonate rocks, ranging from about 170 to 300 km. The powerful argon-argon technique of dating rocks has now confirmed that the crater was formed 65 million years ago when a shallow sea covered the region. Scientists are sure that it was formed by an asteroid about 10 km in diameter zooming at 90,000 km/h.

It is indeed the culprit that wiped out the mighty dinosaurs and with them almost all other species of plants and animals. A few hundred years after this mass extinction, a new world of plants and animals starts to come into being. First shrubs begin to cover the landscape, and then trees begin to grow. It is the start of a new spring.

Back in 1908: How will I bet my proverbial two bobs?

Killer-rocks-from-space scenarios are beloved of astronomers—and science writers—who see Earth-crossing asteroids as a potential threat to humanity.

In this race of dirty dozen suspects, the favourite in a betting shop—run by astronomers—would be a rogue rock from space: not too small to burn up entirely in the upper atmosphere; not too big to leave a hole in the ground. I will bet my proverbial two bobs (two cents, if you prefer) on an asteroid.

Kulik, our beloved investigator, died believing that the rock was big enough to dig a big crater and leave some visible remnants. But we have now a computer ID of the errant rock: a small stony asteroid 30 m across, moving about 54,000 km/h. It exploded 8 km above the Siberian wilderness, leaving only a devastated forest.

No one has found a Chicxulub-like crater in Siberia. The reason is simple: the asteroid was much smaller compared with the 10-km-wide asteroid that wiped out the dinosaurs.

'The problem with the scientists involved in the research is that they get very emotional and less rational about the science,' says Elmar Bartlmae, a German filmmaker who has produced a documentary on the Tunguska event.

For the 'less rational' among us, the story of the Tunguska fireball continues.

Timeline: 115 Years of an Enigma

1908	About 7.15 a.m., 30 June. A fireball explodes in mid-air near the Podkamennaya Tunguska River in Siberia and flattens a vast forest. Nearest eyewitnesses are 70 km from the explosion site, but villagers as far as 700 km away see bright lights in the sky and villagers as far as 1,200 km away hear loud explosions. Siberian newspapers report the explosion, but they are not sure of its nature; some suggest it to be a meteorite. After the explosion, bright 'night glows' are observed in parts of Europe and Asia. This unusual phenomenon is widely reported in newspapers in Britain, Europe and the United States, but no one knows its cause. An observatory 970 km from the explosion site records a magnetic storm that began a few minutes after the blast and lasted for about four hours. Seismic waves are recorded around the world. Six microbarographs in England record air waves created by the blast.
1910	The first expedition to the Tunguska site by a non-Evenk person. A wealthy Russian merchant and goldsmith named Suzdalev is rumoured to have discovered diamonds at the site.
1921	September. Nothing is heard of the explosion until Leonid Kulik, a Russian scientist, is assigned the task of locating and examining meteorites fallen in inhabited regions of Russia before and after the First World War. During his expedition to Siberia, Kulik learns of a

	meteorite that had fallen near the Podkamennaya Tunguska River. The expedition ended without him visiting the explosion site.
1924	S.V. Obruchev, a Soviet geologist, conducts geological studies in the Tunguska region (but not the explosion site).
1925	A.V. Voznesensky, Director of the Irkutsk Magnetic and Meteorological Observatory in 1908, claims that the seismic and air waves recorded by his observatory on 30 June 1908 were both caused by the fall of a giant meteorite.
1926	I.M. Suslov, a Soviet ethnographer, visits Tunguska region. (The famous Suslov crater is named after him.)
1927	21 May. The first Tunguska expedition, led by Kulik, reaches the epicentre of the explosion. Kulik becomes the first scientist to visit the site.
1928	June. Kulik's second expedition reaches the explosion site. Kulik's expeditions are widely reported in British and American newspapers.
1929	Kulik's third expedition. A British scientist notices the coincidence of the date of the Tunguska explosion and the airwaves recorded in England on 30 June 1908.
1930	Other British scientists suggest that the airwaves recorded in England and the remarkable night glows of

	1908 were caused by the Tunguska meteorite.
1934	The British scientist F.J.W. Whipple and the Russian scientist I.S. Astapowitsch independently propose that the Tunguska object was a comet. (Astapowitsch expanded the idea suggested by Academician Vladimir Vernadsky.)
1938	The first aerial photographic survey of the Tunguska region is undertaken.
1939	Kulik's fourth expedition to Tunguska. The last expedition before the start of the Second World War.
1941	The American meteorite expert Lincoln La Paz suggests that the Tunguska object was a contraterrene (antimatter) meteorite.
1942	14 April. Kulik dies in a German prisoner-of-war hospital.
1946	The Russian science-fiction writer Alexander Kazantsev publishes a story suggesting that the Tunguska object was an alien spaceship.
1957	The Russian mineralogist A.A. Yavnel microscopically analyses soil samples brought back by Kulik in 1929 and 1930. These samples were later proved to be of terrestrial origin.

1958	30 June. The Soviet Union releases a commemorative 40-kopeck stamp featuring a portrait of Kulik, to celebrate the 50th anniversary of the Tunguska event.
	The fifth expedition, the first after the Second World War, led by Kirill P. Florenskiy.
	The Interdisciplinary Independent Tunguska Expedition (IITE, known as KSE in Russian) is formed in the Siberian city of Tomsk to discount spaceship theories.
1960	Florenskiy supports the comet theory in an article in the US magazine *Sky & Telescope*. Academician Vassilii Fesenkov also presents arguments in favour of the comet theory to *The New York Times*.
	The Soviet Academy of Sciences builds a simple memorial on Kulik's grave in the town of Spas-Demensk, about 300 m southwest of Moscow.
1961	Florenskiy leads the sixth expedition which continues into 1962.
1964	The Russian science-fiction writers Genrikh Altov and Valentina Zhuravleva suggest that Tunguska was zapped by a laser beam sent by ETs.
1965	The American scientists Willard Libby, Clyde Cowan and C.R. Alturi present a detailed theory showing that the Tunguska object was made of antimatter.
1959	Feliks Zigel, so-called 'Father of Soviet UFOlogy', suggests that the Tunguska object was a UFO.

1966	The English translation of E.L. Krinov's book *Giant Meteorites* is published. Its 141-page section 'The Tunguska Meteorite' has an authoritative account of early research on Tunguska.
1973	The American theoretical physicists A.A. Jackson IV and Michael P. Ryan, Jr say that the Tunguska object was a mini black hole, which passed through Earth and exited through the North Atlantic Ocean.
1976	The first Tunguska book in English is published: *The Fire Came By: The Riddle of the Great Siberian Explosion* by John Baxter and Thomas Atkins.
1975	The Israeli scientist Ari Ben-Menahem concludes that the explosion took place 8.5 km above the ground and had energy of about 12.5 megatons of TNT.
1977	The British scientist Anthony Lawton suggests that the Tunguska fireball was in fact a giant lightning ball.
1978	The Slovak astronomer Lubar Krésak suggests that a piece of comet Encke had exploded at Tunguska.
1983	The American scientist Zdenek Sekanina proposes that the explosion was caused by a stony asteroid. The American scientist Richard Turco suggests that the bright nights were caused by noctilucent clouds produced by the dust that reached the stratosphere. The American chemist Ramachandran Ganapathy says that the globules collected by Florenskiy's 1961-62 expedition are enriched in iridium, a metal that is abundant in extra-terrestrial bodies, and contain other

	evidence of extra-terrestrial origin. He also discovers traces of the Tunguska fireball in an Antarctic ice core.
1984	The Russian scientists Victor Zhuravlev and A.N. Dmitriev present their plasmoid hypothesis.
1989	The first post-Cold War expedition open to international scientists.
1991	The first Italian expedition led by Menotti Galli and Giuseppe Longo. The expedition collects particles from ~~tree~~ resin in Tunguska trees. The particles contain some elements which are commonly associated with stony asteroids. The Russian scientist Andrei Ol'khovatov publishes his 'geometeor' theory. Geometeors are meteor-like luminous objects but of terrestrial origin.
1993	American scientists Christopher Chyba, Kevin Zahnle and Paul Thomas give the asteroid theory new weight and rigour. They say that the explosion released about 15 megatons of energy in the atmosphere at an altitude of about 8 km.
1994	An unknown American writer suggests that the explosion was caused by a Nikola Tesla experiment on a death ray which got out of hand.
1996	The Russian scientist Vladimir Svetsov shows that the entire mass of the Tunguska object vaporised before it could reach the ground. Ablation of the Tunguska debris was total.

1998	Sekanina revisits his asteroid theory and presents new arguments in favour of it. The Russian scientist Vladimir Alekseev suggest that the flight of the object ended in multiple explosions which were responsible for gunfire-like sounds heard by eyewitnesses.
2001	Academician Nikolai Vasilyev, who coordinated the scientific research of 29 Tunguska investigations from 1963 to 2001, dies. A team of Italian scientists, based on an idea of the late Paolo Farinella (1953-2000), calculates 886 valid orbits of the object, of which 83 per cent are asteroid orbits and 17 per cent comet orbits. The German astrophysicist Wolfgang Kundt suggests that the outburst was caused by 10 million tonnes of methane gas which escaped from a volcanic vent.
2002	The Australian physicist Robert Foot suggests that the Tunguska blast was caused by a mirror-matter asteroid.
2004	Dr Vitalii Bronshten, a renowned Tunguska researcher and the main supporter of the comet theory, dies.
2007	After their 2007 on site studies of Lake Cheko, Italian scientists Luca Gasperini, Enrico Bonati and Giuseppe Longo suggest that the lake was formed by the impact of the Tunguska body.
2008	Celebration of the 100th anniversary of the Tunguska event; still there is no definitive answer to the question: what really caused the explosion?

2010	After 2010 expedition to Tunguska, Russian nuclear scientist Vladimir Alexeev says that the blast was caused by a comet.
2013	The Ukraine scientist Victor Kvasnytsya revives the asteroid theory after studying the samples collected from the site of the blast by using new high-resolution analytical an imaging techniques. He claims that the patterning and combination of minerals in the samples are remarkably similar to that in other iron-rich, stony asteroids.
2014	C.D. Froggatt of Glasgow University and H.B. Nelson of The Niels Bohr Institute in Copenhagen propose that that a dark matter ball caused the Tunguska event.
2014	John's Stone: A possible fragment of the Tunguska fireball
2016	The United Nations proclaims 30 June as International Asteroid Day to raise awareness about asteroids and efforts at planetary defences.
2018	At a workshop at NASA Ames Research Center about 50 experts on impact airburst re-examine the Tunguska event and say that Tunguska is the largest cosmic impact witnessed by humans. The Tunguska fireball was a stony (not icy) body, between 164 and 262 feet (50 to 78 metres) in diameter, entering the atmosphere at around 34,000 miles (per hour 54,700 km) per hour and exploded at an altitude of 6 to 9 miles (9.6 to 14.5 km).

2023	NASA reiterates that an asteroid plunged into Earth's atmosphere and exploded in the sky over Siberia, and it continues to search for and monitor near earth objects (NEOs) that come within 30 million miles (48.28 million km) of Earth's orbit.

www.ingramcontent.com/pod-product-compliance
Lightning Source LLC
Chambersburg PA
CBHW071042240526
45471CB00014B/308